STARLORE
OF THE
CONSTELLATIONS

Starlore of the Constellations

Geoffrey Cornelius

First published in the UK in 2000 by Duncan Baird Publishers

This edition published in the UK and USA in 2024
by Watkins, an imprint of Watkins Media Limited
Unit 11, Shepperton House, 83–89 Shepperton Road
London N1 3DF

enquiries@watkinspublishing.com

Illustrations by Emma Harding and Alice Claire Coleman

A CIP record for this book is available from the British Library

ISBN: 978-1-78678-924-2 (Paperback)
ISBN: 978-1-78678-925-9 (eBook)

10 9 8 7 6 5 4 3 2 1

Printed in China

www.watkinspublishing.com

The Astronomy, Myth and Symbolism of the Night Sky

STARLORE
OF THE
CONSTELLATIONS

GEOFFREY CORNELIUS

CONTENTS

HOW TO USE THIS BOOK

The Night Sky

The first section of the book details the historical and astronomical background to our understanding of the heavens. It covers the principles of sky-watching, introducing the concept of the Celestial Sphere, the zodiac and the ecliptic.

The Full-Sky Maps

These maps, divided into two sections for Northern (30°N) and Southern (45°S) hemispheres, show the movement of the constellations through the year.

The Major Constellations

Using commissioned artwork (see key, page 9), 40 of the best-known constellations are given detailed treatment in alphabetical order of the full Latin name, following the internationally-recognized astronomical convention. Underneath the main heading is the standard three-letter abbreviation (for example, for Taurus this is Tau), the genitive form of the name (often used in star names, meaning "of" the constellation; δ Tauri is the δ-star of Taurus), and the popular name or translation (The Bull).

Each constellation map shows stars to magnitude 5.25 (see pages 20–1), the limit for naked-eye observation. Stars are given Greek letter (Bayer) designations (see complete alphabet on page 8) or numerical (Flamsteed) designations. Some dimmer stars of exceptional interest are also shown. α is usually the brightest star, β the next-brightest, and so on.

Each constellation is introduced from the point of view of practical observation: how to recognize it, the limits of observation, and its date of culmination at midnight (1am when Daylight Saving or summertime is used). The major stars for each constellation are tabulated giving the star's name, its magnitude and, in most cases, its colour. Any interesting deep-sky objects such as nebulas or supernovas (see page 25) are also listed. Finally, a description is given of the lore associated with the constellation.

With some of the Major Constellations, small "pointer" maps are used to help you find your way around the night sky (see, for example, page 53). There are also three main Signpost Maps in the book to help you to locate the circumpolar constellations for the North and South Poles (see pages 126 and 162) and the Andromeda Group (see page 72).

The Minor Constellations

The remaining 48 constellations are given two-colour maps (see pages 135–65); their treatment follows the Major Constellations, but in condensed form.

The Wandering Stars

This section (see pages 167–77) describes the naming and myth of the Sun and Moon and the planets of our Solar System.

The Appendices

The Zodiac Signs Table (see page 180) shows the the date that the Sun enters each 30° section (sign) of the zodiac and the degrees of celestial longitude at which the sign begins.

The Star Tables (see pages 181–7) provide an alphabetical index to the named stars in the book, listing in each case the page number on which the star can be found, its designation, constellation, and full celestial coordinates.

The Glossary (see pages 188–90) defines the major technical terms used in the book.

An Index appears on pages 193–200.

Image from Bridgeman Art Library
/ Musée Condé, Chantilly

Detail from a celestial map of the Southern-hemisphere sky (c.1650), by Andreas Cellarius. Bridgeman Art Library/ British Library.

STAR MAGNITUDES

- ●   O
- ●   1
- ●   2
- ●   3
- ●   4
- ●   5

THE GREEK ALPHABET

α	alpha
β	beta
γ	gamma
δ	delta
ε	epsilon
ζ	zeta
η	eta
θ	theta
ι	iota
κ	kappa
λ	lambda
μ	mu
ν	nu
ξ	xi
ο	omicron
π	pi
ρ	rho
σ	sigma
τ	tau
υ	upsilon
φ	phi
χ	chi
ψ	psi
ω	omega

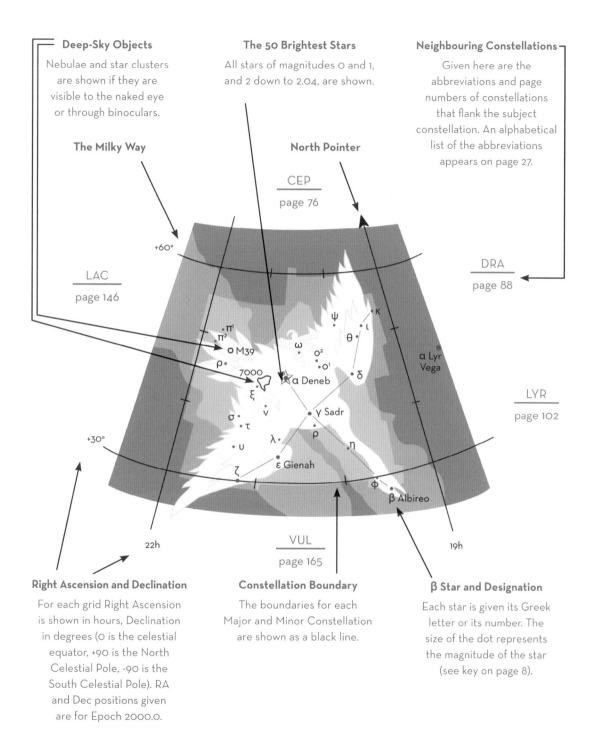

Deep-Sky Objects

Nebulae and star clusters are shown if they are visible to the naked eye or through binoculars.

The Milky Way

The 50 Brightest Stars

All stars of magnitudes 0 and 1, and 2 down to 2.04, are shown.

North Pointer

Neighbouring Constellations

Given here are the abbreviations and page numbers of constellations that flank the subject constellation. An alphabetical list of the abbreviations appears on page 27.

CEP
page 76

LAC
page 146

DRA
page 88

+60°

ψ
κ
ι
θ
π¹
π²
ω
ο²
δ
α Lyr
Vega
ρ
ο M39
ο¹
7000
☆ α Deneb
ξ
LYR
page 102
ν
γ Sadr
σ
τ
ρ
λ
η
υ
ζ
ε Gienah
φ
β Albireo

+30°

22h

VUL
page 165

19h

Right Ascension and Declination

For each grid Right Ascension is shown in hours, Declination in degrees (0 is the celestial equator, +90 is the North Celestial Pole, -90 is the South Celestial Pole). RA and Dec positions given are for Epoch 2000.0.

Constellation Boundary

The boundaries for each Major and Minor Constellation are shown as a black line.

β Star and Designation

Each star is given its Greek letter or its number. The size of the dot represents the magnitude of the star (see key on page 8).

A 17th-century map of the Northern- and Southern-hemisphere skies. Also shown are (clockwise from top left) Tycho Brahe's solar system, Ptolemy's system, the motion of the Moon, the Earth's motion around the Sun, Copernicus' solar system (shown too as a detail, below) and the Moon's oppositions.

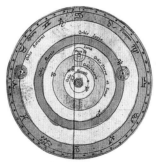

A representation of night from a pair of paintings entitled *Day and Night*, by Edward Robert Hughes (1841–1914). The crown of stars around the figure's head is reminiscent of the constellations Corona Borealis and Corona Australis, the Northern and Southern Crowns (see pages 80–3).

THE NIGHT SKY

For of celestial bodies first the Sun / A mighty sphere he framed, unlightsome first, / Though of Ethereal mould: then formed the Moon / Globose, and every magnitude of stars, / And sowed with stars the heaven thick as a field. "

John Milton, Paradise Lost; Book VII, pages 354–358 (1667)

For the ancients the contemplation of the heavens was the noblest of all sciences. Modern astronomy itself has grown up within the poetic vision of the cosmos that has come down to us from Mesopotamia and Egypt and through ancient Greece. The result has been the creation of a universal stellar imagery, anchored in Greek myth and overlaid with Arabic learning, yet still a part of the worldwide currency of modern culture. The following pages unfold the essence of a marvellous tradition that combines archaic myth with modern science.

THE ANCIENT SKIES

The identification of stars in groups is found among all cultures at all stages of development. Although many groupings belong to the perceptions of a particular society, curious parallels often occur. According to the astronomer Julius Staal (1917–1986), it was a widespread tradition among the Native American Indians to see a bear in α, β, λ and δ Ursae Majoris (the Greater Bear; page 124). The three stars of the handle of the Plough or Big Dipper, part of Ursa Major (ε, ζ and η), were seen as three hunters, whereas for classical Greek tradition these form the bear's tail. The bear symbolism can be traced to early origins on the Indian subcontinent, where the seven stars of the Plough or Big Dipper were the seven *Rishis*, or sages; *rishi* has a Sanskrit root meaning "bear".

Another recurring myth-complex of this asterism is its association with a wagon or carriage, found in Babylonian representations, and also in ancient China (see page 14).

But what was the original impetus behind our mapping and naming of the star-fields? Firstly, ancient calendars are lunar, rather than solar, and in all probability it was a desire to track the Moon that led to the first identifications of significant groups of stars. A widespread early development was the tabulation of lunar mansions; these were known in Arabic as *al-manazil*, in India as *nakshatra*, in Hebrew as *mazzaloth*, and in China as *hsiu*. The mansions are star-groups or regions along the ecliptic,

or in ancient China the equator, by which lunar motion may be tracked. By dividing the path of the Moon across the stars into 28 (sometimes 27) sections, it will be seen to cross into a new mansion each night.

A second fundamental element of observation is the apparent daily rotation of the sky. An understanding of this was well developed among the greatest of all ancient sky-watchers, the priests of the Assyro-Babylonian culture, in the fertile region of Mesopotamia. In the omen-texts known as *Ea Anu Enlil* (1400–1000 BCE), we find the heavens divided into three "roads" dedicated to the three gods. Ea took the outer road, traversed by the stars south of the equator; his son Enlil was given the inner road of the circumpolar stars; while Anu held the middle road, around the equator. Along each road, 12 star-gods showed the months of the year by their heliacal risings (their first appearances ahead of the Sun, after a period of being unseen), and at any one time 18 of the stars would be visible.

From the 6th century BCE, Greek civilization assimilated Mesopotamian, Persian and Egyptian astronomy, astrology and mythology. Remnants of the equatorial system of the *Ea Anu Enlil* became fused into the distinct development of the ecliptic-based zodiac (see pages 17–19) around the 5th century BCE, and this was the historical basis of all later Western astrology and astronomy. The Greek zodiac also incorporated a key element of earlier lunar observations, as it was a band 6° either side of the Sun's path (the *ecliptic*), covering the maximum swing of the Moon above and below that path.

The culmination of these endeavours is found in the works of Claudius Ptolemy (2nd century CE). Ptolemy, reworking existing data (especially the 2nd-century-BCE observations of the astronomer Hipparchus), produced a catalogue of more than 1,000 stars visible from Mediterranean lands. He grouped them into 48 constellations, comprising the 12 zodiacal figures, 21 constellations to the north, and 15 to

the south. Ptolemy's catalogue served as the definitive authority for nearly one and a half millennia. It was not until the period of European expansion, especially from the 15th century onward, that we find any major additions or amendments, most notably in the mapping of the Southern constellations. An astronomical convention in 1930 precisely established the constellation boundaries; henceforth any star could be ascribed its proper constellation without confusion. This universal system leaves the Greek stellar figures almost untouched by centuries of celestial exploration.

The zodiac on the ceiling of the great temple at Dendera, Egypt. This illustrates
a transitional stage, from the 3rd century BCE, in the development of the
zodiac in Egypt, fusing Mesopotamian and ancient Greek motifs.

A celestial bureaucrat in his carriage, representing the Plough or Big Dipper, in a relief from the Wu Liang tomb, China (2nd century CE). The figure is reversed as if seen from the outside of the Celestial Sphere. Close by Mizar (ζ UMa) in the right half of the picture, the star held up by a spirit is probably Alcor (80 UMa).

THE MOTION OF THE STARS

The *Celestial Sphere*, a projection of the Earth's surface into the sky with the Earth at its centre, is a fundamental concept in astronomy. Lines of sight are projected onto it, and measurements of arc between celestial objects are made on the inside surface of this imaginary globe.

At any position on the Earth a horizon circle cuts the Sphere into upper (visible) and lower (invisible) sections. The *astronomical horizon* is any horizon circle whose plane cuts through the centre of the Sphere. We see a close approximation to the notional astronomical horizon at sea and in flat deserts. The point on the Celestial Sphere directly overhead is the *zenith*. At the zenith's opposite point on the Celestial Sphere lies the *nadir*. A measure above the horizon is termed *elevation*; and the zenith is by definition at the maximum elevation of 90°, a quarter circle, above the horizon.

All the stars and planets lying on the Sphere appear to rotate around the Earth once every 24 hours. From modern science we know that it is the Earth's turn on its own polar axis that makes the sky "turn". If this polar axis is projected onto the Celestial Sphere, it pierces it at the *Celestial Poles*: the North Celestial Pole is the zenith for an observer at the Earth's North Pole; the South Celestial Pole is

The Celestial Sphere

The projection of the Earth's poles onto the Sphere gives Celestial North and South, with the celestial equator midway. The ecliptic (tilted at 231/2° to the equator) marks the Earth–Sun orbital plane. Equator and ecliptic intersect at the March equinox point (0° Aries, the starting point for celestial measurement), and at the opposite point on the Sphere (the September equinox). Celestial objects are measured against the equator (Right Ascension and declination) or the ecliptic (celestial longitude and latitude).

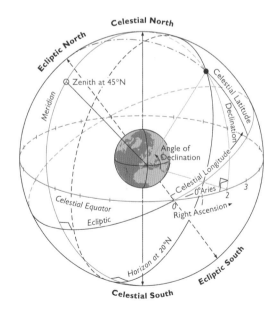

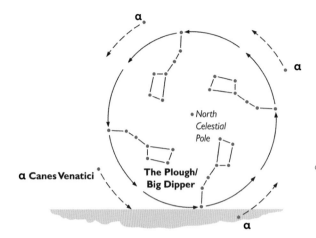

α

α

α

North
Celestial
Pole

α Canes Venatici

**The Plough/
Big Dipper**

α

Circumpolar Motion

As we move toward the pole, stars that rise and set at the equator become circumpolar, ie. they never set. From 41°N the Plough becomes entirely circumpolar, because all of its stars have declination greater than 49°N (90°–41° = 49°). The star Cor Caroli (α Canes Venatici) is at declination 38°N, so at 41°N geographic latitude it passes daily below the horizon. At the pole the whole hemisphere becomes circumpolar.

the zenith at the Earth's South Pole. Midway between the Celestial Poles is the *celestial equator*. During the night, the stars cross the sky above the horizon in circles parallel to the celestial equator.

The location of any celestial object, including the stars and planets, can be pinpointed on the Celestial Sphere in two ways. The system of Right Ascension (RA) and declination is equatorial. RA is measured in 24-hourly sections along the equator; declination is measured in degrees above or below the equator (0° to +90° in the North; 0° to −90° in the South). The system of the ecliptic (the circle made by the apparent path of the Sun) measures the positions of stars in celestial longitude (0° to 360° along the ecliptic) and celestial latitude (0° to 90°) north (+) or south (−) of the circle. Both systems begin at 0° Aries (see page 17).

The observer's location on the Earth changes the way in which the Sun, Moon and stars appear to rise and set. At the equator the Sphere appears to turn so that stars rise at right angles to the horizon; north or south of the equator the angle of rising and setting becomes increasingly slanted. Hence the short twilight and dawn on and near the equator, and the long, slantwise rising and setting of the Sun at latitudes toward the poles.

The line that joins the North and South Poles and runs through the observer's zenith is termed the *meridian*. This cuts the horizon at right angles to the cardinal points north and south. The meridian is a key to astronomical observation and to time-keeping. As a rising star crosses the upper meridian (that part of the meridian above our heads), it is said to *culminate*. The culmination of the Sun — sundial noon — is the basis of the ancient measurement of time.

THE ZODIAC AND PRECESSION

The Earth rotates at a tilt to its plane of orbit, and in consequence the celestial equator (a projection of the terrestrial equator) is tilted with respect to the ecliptic (see page 15) at 23° 26'. For one half of the year the Sun is north of the equator, for the other half it is south, and this variation produces changes in the lengths of day and night, and in the seasons.

The ecliptic and celestial equator cross at two points 180° apart: these points are called the *equinoxes*. When the Sun arrives at either of these points, day and night are of equal lengths worldwide.

After the March equinox, the Sun moves north of the equator until it reaches a maximum north declination of 23° 26' at the *solstice* ("stand-still"; when the Sun is at its highest and lowest positions in the sky for the year) around June 22. It then "drops" toward the September equinox, crosses south of the equator, and moves toward its maximum south declination of 23° 26' at the opposite solstice around December 22.

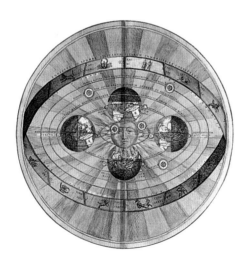

A 16th-century band of the zodiac around the Copernican motion of the planets. The Earth is shown in four positions along its orbit.

The *zodiac* is a band of sky lying 6° either side of the ecliptic, its width reflecting the maximum celestial latitude reached by the Moon. The zodiac is divided into twelve 30° *signs*, beginning at 0° Aries, the March equinox. The fourth sign, Cancer, begins at 90° from the starting point of the zodiac. When the Sun reaches this point, it is the June solstice. Similarly, the Sun's ingress into Capricorn marks the December solstice.

The zodiac is a means of measuring the Sun's motion through the year. The signs were originally identified with the imagery of the constellations that lay behind them on the Celestial Sphere. However, it is important to distinguish the 12 equal signs (functions of the Earth-Sun orbital cycle) from the irregularly-sized, fixed-star constellations from which both their names and symbolism are derived.

In the closing centuries BCE, the March equinox point fell at the border of the constellations Aries and Pisces. The 30° of the sign Aries thus fell roughly over the fixed stars of the constellation Aries. However, this framework undergoes a small but cumulative change termed *precession* due to the slow rotation of the Earth's polar axis around the ecliptic pole (like a gyroscope) over 25,868 years. Each year, as the Sun returns to 0° Aries, its position against the background stars will have slipped back 50" of arc — 1° in 72 years. At the dawning of the 3rd millennium, this motion has produced a discrepancy of nearly a whole constellation; our modern sign Aries now falls over the stars of Pisces. This leads to the symbolism of the Great Ages. As we are now in the closing era of the Age of Pisces, we will soon follow the March equinox point backward into the Age of Aquarius.

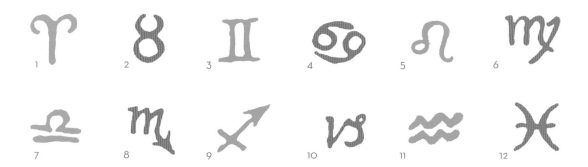

Set out above are the zodiacal sigils (signs), in their order around the band of the zodiac.
They appear as follows: 1 Aries; 2 Taurus; 3 Gemini; 4 Cancer; 5 Leo; 6 Virgo;
7 Libra; 8 Scorpio; 9 Sagittarius; 10 Capricorn; 11 Aquarius; 12 Pisces.

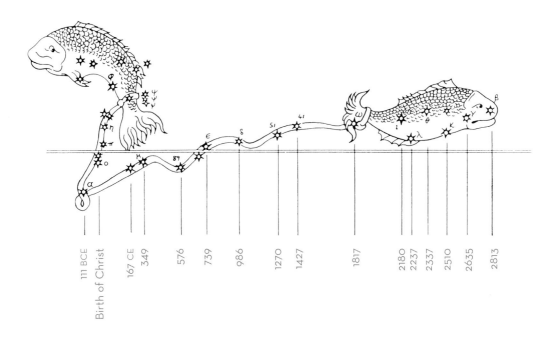

111 BCE

Birth of Christ

167 CE
349
576
739
986
1270
1427
1817
2180
2237
2337
2510
2635
2813

Precession (the movement of the March equinox point – 0° Aries – backward through the fixed stars over a period of 25,868 years) is caused by the slow rotation of the Earth's polar axis. The dates given here show the conjunctions of 0° Aries with the stars of Pisces over three millennia.

MAGNITUDE, DISTANCE AND COLOUR

In the 2nd century BCE, Hipparchus, a Greek astronomer listed roughly 850 visible stars, grouping them into brightness ranges of *apparent magnitude* (the light that a star presents to the naked eye). The brightest stars became first magnitude, the next second magnitude ... down to a dim sixth magnitude.

This method served well until the development, early in the 17th century, of the telescope, which uncovered a rich crop of previously invisible stars, all in need of accurate cataloguing. Hipparchus' approach could neither cope with the new volume of stars, nor allow for a distinction between the very brightest, grouped together in the first magnitude with greatly varied luminosities.

The system was finally calibrated in its modern form by the English astronomer Norman Tobert Pogson in 1856. He established the ratio between magnitudes as 2.512, so that a five-magnitude difference is the same as a one hundredfold change in brightness (2.512 to the power of 5). To classify more stringently the very brightest stars (those that previously had been magnitude 1), Pogson's system gives the most luminous stars negative numbers (Sirius is the brightest at magnitude −1.46). This extension allows the planets, frequently much brighter than the brightest stars, to be classified too (the brilliant Venus shines out at −4.4; see page 172).

Each magnitude range, apart from the first, is demarcated by the half-way point between two magnitudes; so, a star will be second magnitude if its precise magnitude value lies between 1.50 and 2.49. As a recognized convention of reference, any star brighter than magnitude 1.5 is termed first magnitude.

Stars of the fourth and fifth magnitude are distinctly faint, and many fifth-magnitude stars

The brightness (vertical axis) and temperature (horizontal) of stars (shown by dots), from the hottest (blue) to the coolest (red); white stars (omitted here) occur between blue and yellow.

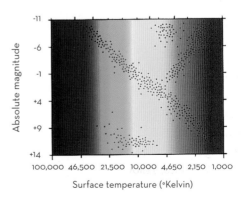

are indistinguishable unless viewing conditions are very good. In principle, some 3,000 stars (of magnitude less than 5.50) are available for naked-eye observation; stars of the sixth magnitude require exceptional eyesight.

The figures for apparent magnitude are *zenith* observations (see pages 15–16) – they relate to the brightness of the star when it is directly overhead. For practical purposes a star will hold its full magnitude from halfway up in the sky toward the zenith. Closer to the horizon, stars appear dimmer because their light is absorbed on its long slantwise path through the Earth's atmosphere. As a useful rule-of-thumb, at 10° above the horizon a star or planet drops one magnitude and at 4° it drops two magnitudes.

Apparent magnitude, however, gives us no indication of how much light is poured out by the star at source – its *absolute magnitude*, the brightness that a star would appear to have (the apparent magnitude) if it were 30.26 light years from Earth (see diagram, right). By way of comparison, if the Sun were 30.26 light years away, it would appear to be of magnitude 4.8, and this is therefore its absolute magnitude.

There is a more direct relationship between astrophysics and naked-eye observation in the colours of the stars. The distinct hues reflect surface temperatures. Blue stars burn at the highest temperature (as high as 1,280,000°F / 40,000°C), while red stars are slow-burning and relatively cool (as low as 96,000°F / 3,000°C). For naked-eye purposes, the hues run in the following order (hottest to coolest): blue, blue-white, white, yellow-white, orange, red.

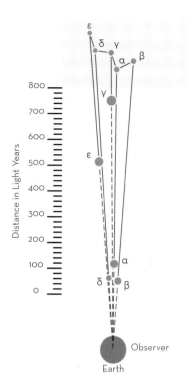

The Celestial W of Cassiopeia

The distances of the stars in light years (the distance that light travels in one year, at 186,000 miles/sec or 300,000 km/sec) are shown along the vertical axis. The dot sizes in the lower part of the diagram are the *absolute magnitudes* of the stars; those in the upper part are *apparent magnitudes*. Four of the stars (α, β, γ, δ) are close in apparent magnitude, all lying between 2.3 and 2.7. The nearest of the group, Caph (β), is 46 light years distant, with an absolute magnitude of 1.4. Schedar (α) has roughly the same apparent magnitude as β but approaches two-and-a-half times the distance from Earth, and so must be brighter at source (absolute magnitude −0.9); while Cih (γ), again similar to the other two in apparent magnitude, is 17 times the distance (approx. 780 light years), and so has an even greater absolute magnitude (−4.6).

THE SUN,
MOON AND PLANETS

The *Solar System* is the heliocentric (Sun-centred) organization of the planets first published by Copernicus in 1543.

The Sun is believed to have been created in an eruption of galactic energy (a *supernova*) some 4,000 million years ago. Fragments from the explosion became the planets. Composed mainly of hydrogen and helium gas, the Sun is 870,000 miles (1.4 million km) in diameter. Its heart, which resembles a vast nuclear reactor, has a temperature of 27 million°F / 15 million°C, fusing hydrogen into helium and releasing huge amounts of radiation.

The Sun's surface temperature is around 10,500°F / 5,800°C. Cooler patches appear as darker *sunspots*. These can reach diameters of more than 60,000 miles / 100,000 km, making them visible to the naked eye (**however, never look at the Sun, either directly or through a mirror**). In terms of absolute magnitude (see page 21), the Sun is a modest 4.8, compared with, say, Rigel (β Ori), at –7.0.

The Moon is a natural satellite unique in its relatively large size compared to its parent body, with a diameter of 2,160 miles / 3,476 km, roughly a quarter that of the Earth. It spins on its axis once in 27$\frac{1}{3}$ days, the same amount of time that it takes to orbit the Earth. This *captured rotation* means that the same face of the Moon is always turned toward us.

The Moon's phases are produced by the changing angle by which the Sun's light is reflected on the Moon's face. When Full, the Moon is on the other side of the Earth from the Sun, directly opposite to it, and the light of the Sun is reflected from the Moon's whole hemisphere. The complete cycle of phases averages 29$\frac{1}{2}$ days (the time it takes for the Moon to return to the same position along its orbit, from a position on Earth).

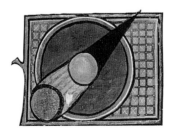

A solar eclipse shown in a manuscript illumination from
L'ymage du monde, dated 1245, by Gautier de Metz.

The distance of the Moon from the Earth (just 239,000 miles / 384,000 km) makes its disk equal to the larger but more distant Sun (at 93 million miles / 150 million km). Culturally, this pure coincidence has enhanced the pairing of the Sun and Moon as symbolic equals (see pages 168–70). One observational effect is that at a *total solar eclipse* the Sun's disk can be briefly but wholly covered by the disk of the Moon. This can occur only at the time of a New Moon, when the satellite falls directly on a line of sight between the Earth and the Sun from a particular position on the Earth's surface. Sometimes small variations in the Moon's orbit reduce the size of its disk, in which case a ring of sunlight glows around the edge, causing an *annular eclipse*. From any position on Earth lying beyond the narrow eclipse path (the band of shadow that falls across the Earth's surface), the Moon will not appear fully to cover the Sun. Most often an eclipse is not visible as total anywhere on the Earth, and only a section of the Sun is hidden (a *partial solar eclipse*).

A *lunar eclipse* occurs when there is a Full Moon, and the long cone of the Earth's shadow, or *umbra*, strikes exactly on the Moon's surface.

So, an eclipse can occur only when the Earth, Sun and Moon are on the same line of sight and the same plane — the plane of the *ecliptic*.

The ecliptic is the great circle that is described by the Earth as it orbits the Sun, and it is therefore also the apparent track of the Sun through the fixed-star background. The astronomical importance of the ecliptic relates not only to eclipses, but also to the other planets. The Solar System is like a flat dish, because most of the planets orbit the Sun within a few degrees of tilt compared with the plane of the ecliptic. Only Pluto has a distinctly marked orbital inclination of 17°. Locating this great circle with respect to the constellations makes it easier for us to identify the planets in the sky.

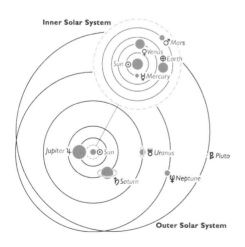

Inner Solar System

Mars · Venus · Earth · Sun · Mercury · Jupiter · Sun · Uranus · Pluto · Saturn · Neptune

Outer Solar System

The Solar System
This diagram shows the relative positions and sizes of the planets, with the Inner Solar System (the planets within the asteroid belt) magnified seven times in relation to the Outer Solar System. Each planet is shown with its sigil (conventional sign). The Polish astronomer Copernicus published the *heliocentric* (Sun-centred) view of the universe in 1543. The first hypothesis for this came from Aristarchus in 330 BCE, at a time when the universe was thought to be *geocentric* (Earth-centred).

METEORS AND COMETS

Historically regarded as heavenly portents, millions of comets are known to exist on the remote outer edges of the Solar System. Some enter greatly elongated orbits around the Sun and return periodically into view.

The best-known comet is that of Halley, who in 1705 computed its orbit to have a period of 76 years. Its appearance in 1066 was seen as an omen for the invasion of England by William of Normandy. It last appeared in 1985–6.

Meteors occur when the Earth moves through the trail of dust and gas left in the wake of a comet. Particles of debris streak into the upper atmosphere, momentarily appearing as magnitude 2 or 3, but sometimes reaching magnitude 1, in which case they are *fireballs*, able to outshine even the brightest stars.

When the Earth passes through a dense zone of particles, it is possible to see a meteor shower. Many of these reappear each year as the Earth returns in its orbit to the same dusty region. They appear to fan out from a central point: the *radiant*. The strength of a shower is designated by its Zenithal Hourly Rate (ZHR): the theoretical number of meteors observable, brighter than magnitude 6.5, under ideal conditions with the radiant directly overhead.

Comet Ikeya-Seki
The brightest comets have glowing heads and immense tails, visible for months at a time.

MAJOR METEOR SHOWERS

Meteor Shower	Limits	Best Dates	ZHR	Constellation
Quadrantids	Jan 1–6	Jan 3–4	100	Boötes
Eta Aquarids	May 1–10	May 5–6	35	Aquarius
Delta Aquarids	July 15–Aug 15	July 28–9	20	Aquarius
Perseids	July 23–Aug 20	Aug 12–13	75	Perseus
Orionids	Oct 16–27	Oct 22	25	Orion
Taurids	Oct 20–Nov 30	Nov 4	10	Taurus
Leonids	Nov 15–20	Nov 17–18	10	Leo
Geminids	Dec 7–15	Dec 13–14	75	Gemini

DEEP-SKY OBJECTS

It is estimated that around 15 per cent of the mass of our Galaxy is made up of *nebulae* (from the Latin *nebula*, meaning "cloud"): immense clouds of gas and dust that periodically coalesce and heat up, generating nuclear reactions when stars are born. The Orion Nebula (M42) is the best known of these formations, clearly visible to the naked eye and marking the hunter's dagger (see pages 106–7).

A *planetary nebula* is a minute variant of a nebula; the term refers to the shell of gas given off by some stars toward the ends of their lives. This sometimes gives a ring shape to a star, reminiscent of the rings that appear around the planets Saturn and Uranus.

In addition, at much greater distances than most single stars, it is possible to see *globular clusters*; around a hundred have so far been observed. These are among the oldest formations in the Galaxy. They are held together by their own gravitational fields and comprise anything from 100,000 to several million stars. A few globular clusters can be seen with the naked eye. The brightest is NGC 5139, ω Centauri, at a distance of 17,000 light years; NGC 104, 47 Tucanae, also in the Southern hemisphere, comes a close second. The Northern hemisphere observer should be able to spot NGC 6205 (M13) in Hercules (see pages 94–5), especially with the aid of binoculars.

Around one thousand *open clusters* have also been identified. These are much looser groupings of recently formed stars, often still surrounded by wisps of gas. They comprise anything from a few dozens of stars to several hundreds. The Pleiades (M45) and the Hyades, both in Taurus, are among the best-known examples (see pages 122–3).

The final destruction of one of the more massive stars produces a *supernova*, like an immense nuclear explosion. The gas ejected from the explosion draws into its train other interstellar nebula material, creating a vast glowing remnant. Supernovae can flare in our skies as brilliant "new stars", but these will die away within the space of a generation. The best-known supernova is the Crab Nebula in Taurus (see page 120).

An ultraviolet image of NGC 5139, ω Centauri, in the constellation Centaurus. This globular cluster was first recorded in 1677 by Edmund Halley.

WORKING WITH THE STARS

To appreciate the stars in their full glory, we need good conditions — even the light of the Full Moon chases minor stars from view. Anyone who has contemplated the skies in clear desert air or on a frosty night deep in the countryside soon realizes the disadvantage faced by the town-dweller from the glow of streetlights.

It takes time for our vision to adapt to the night sky after coming out of the light. Once this is achieved a useful aid for reading star maps is a torch with a red light — this will not readjust your eyes, as would a white light. Binoculars are a marvellous adjunct, enabling us to make out up to 10 times as many stars. The key to choosing a good pair of binoculars is not magnification but light-gathering power, field of view and lightness of use. **Remember, when using any form of magnification, *never* direct it at or near the Sun — irreparable damage to the eyes can occur in an instant.**

Astronomers observing the stars at the Galatea Tower in Constantinople; from a 16th-century miniature.

Effective observation of stars and planets depends firstly on being able to find the Major Constellations by locating a few of their brightest stars. We also need some understanding of the motions of the Celestial Sphere and the Sun, as described on previous pages. Among the most obvious considerations is whether a constellation can be seen at all from the observer's geographic latitude; this will depend upon its *declination* (see page 16). From a given location on the Earth, a star will be circumpolar (always visible) if its declination is greater than the geographic latitude of the place. Conversely a star is lost for the Earth's opposite hemisphere at geographic latitudes greater (nearer the pole) than 90° minus the star's declination. The brilliant Southern star Canopus (α Car) is at declination 52°S 42'. It is invisible at geographic latitudes north of 37° 18' (90° − 52° 42'). There is no point in seeking Canopus from New York (40°N 40'), but at the right hour at the right time of year we should spot it over the horizon at Miami (25°N 45').

For the astronomer, the turning of the heavens is tracked in sidereal time, which measures hours not from the position of the Sun, but from successive culminations of the March equinox point, 0° Aries. Thus 24 sidereal hours (one sidereal day) are measured from the moment when 0° Aries culminates to its culmination on the following day.

A useful tool for stellar observation is the *planisphere*, a full-sky map with a revolving "window" showing the horizon line for a particular latitude. It shows which constellations are visible at any given date and time.

A good starting point for finding the stars are the maps on pages 28—39, giving the orientation of the heavens from 30°N in the Northern hemisphere and 45°S in the Southern, at two-monthly intervals. Each map has a timetable for observation, giving date, Local Mean Time (LMT) and Daylight Saving Time (DST).

Abbreviations of the Constellation Names

And	Andromeda	**Cyg**	Cygnus	**Pav**	Pavo
Ant	Antlia	**Del**	Delphinus	**Peg**	Pegasus
Aps	Apus	**Dor**	Dorado	**Per**	Perseus
Aql	Aquila	**Dra**	Draco	**Phe**	Phoenix
Aqr	Aquarius	**Equ**	Equuleus	**Pic**	Pictor
Ara	Ara	**Eri**	Eridanus	**PsA**	Piscis Austrinus
Ari	Aries	**For**	Fornax	**Psc**	Pisces
Aur	Auriga	**Gem**	Gemini	**Pup**	Puppis
Boö	Boötes	**Gru**	Grus	**Pyx**	Pyxis
Cae	Caelum	**Her**	Herucles	**Ret**	Reticulum
Cam	Camelopardalis	**Hor**	Horologium	**Scl**	Sculptor
Cap	Capricornus	**Hya**	Hydra	**Sco**	Scorpius
Cep	Cepheus	**Hyi**	Hydrus	**Sct**	Scutum
Car	Carina	**Ind**	Indus	**Sex**	Sextans
Cas	Cassiopeia	**Lac**	Lacerta	**Ser**	Serpens
Cen	Centaurus	**Leo**	Leo	**Sge**	Sagitta
Cet	Cetus	**Lep**	Lepus	**Sgr**	Sagittarius
Cha	Chamaeleon	**Lib**	Libra	**Tau**	Taurus
Cir	Circinus	**LMi**	Leo Minor	**Tel**	Telescopium
CMa	Canis Major	**Lup**	Lupus	**TrA**	Triangulum Australe
CMi	Canis Minor	**Lyn**	Lynx	**Tri**	Triangulum
Cnc	Cancer	**Lyr**	Lyra	**Tuc**	Tucana
Col	Columba	**Men**	Mensa	**UMa**	Ursa Major
Com	Coma Berenices	**Mic**	Microscopium	**UMi**	Ursa Minor
CrA	Corona Australis	**Mon**	Monoceros	**Vel**	Vela
CrB	Corona Borealis	**Mus**	Musca	**Vir**	Virgo
Crt	Crater	**Nor**	Norma	**Vol**	Volans
Cru	Crux	**Oct**	Octans	**Vul**	Vulpecula
Crv	Corvus	**Oph**	Ophiuchus		
CVn	Canes Venatici	**Ori**	Orion		

These abbreviations are used in the full-sky maps on pages 28—39, and for the neighbouring constellations in the star maps on pages 42—165.

NORTHERN HEMISPHERE

If we look south, the giant Orion (Ori) appears in the centre of the sky. His belt of three stars creates a line toward the southeast, pointing to Sirius in Canis Major (CMa). The line of the belt northwest goes to Aldebaran in Taurus (Tau). Over the giant's head to the north lies beautiful Capella in Auriga (Aur). Betelgeuse, on Orion's eastern shoulder, is one corner of the Winter Triangle; find Sirius to form another corner of the triangle and complete the shape to the northeast with Procyon in Canis Minor (CMi). Rigel in Orion's western knee is the giant's brightest star; compare its blue-white colour to the red of Betelgeuse

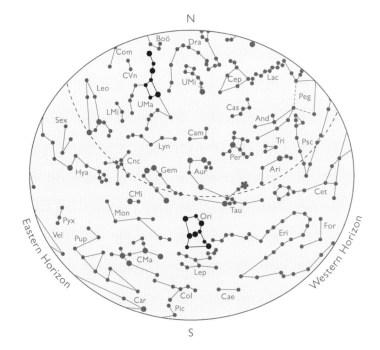

Timetable	Date	LMT	DST
	January 1	11pm	Midnight
	January 15	**10pm**	**11pm**
	February 1	9pm	10pm
	February 15	8pm	9pm

For key to constellation abbreviations, see page 27.

NORTHERN HEMISPHERE

In the centre of the sky, the sickle shape of Leo's head curves up from bright Regulus. Far west of Leo lies Procyon in Canis Minor (CMi), beneath the Gemini twins (Gem). Farther west Orion (Ori) sets, soon to be lost from view; but in the northwest bright Capella in Auriga (Aur) can still be seen. North of Leo is the circumpolar realm and the instantly recognized Plough (Big Dipper), part of Ursa Major (UMa). Join the two right-hand stars of this asterism for a line that points to Polaris, the North Pole star, in Ursa Minor (UMi). Between Gemini and Leo, in Cancer (Cnc), is the Beehive Cluster, M44, visible with binoculars.

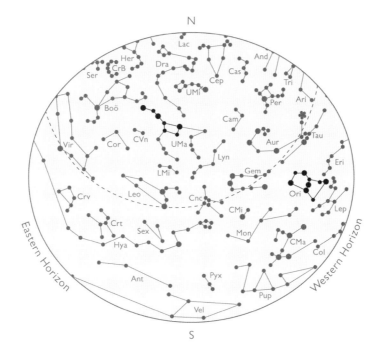

Timetable	Date	LMT	DST
	March 1	11pm	Midnight
	March 15	**10pm**	**11pm**
	April 1	9pm	10pm
	April 15	8pm	9pm

For key to constellation abbreviations, see page 27.

THE NIGHT SKY

NORTHERN HEMISPHERE

To the north, the seven bright stars of the Plough or Big Dipper, part of Ursa Major (UMa), provide a reference frame for spring and early-summer skies; it is easy to follow the handle of the asterism to Arcturus in Boötes (Boö) and Spica in Virgo (Vir), over the southern horizon (see also page 53). To the west of Virgo along the ecliptic, Leo is readily discerned; to the east lies Libra (Lib). Northeast of Boötes, beyond Hercules (Her), Lyra's (Lyr) bright star Vega dominates the sky. The faint Hydra (Hya), the watersnake, can just be made out, its tail immediately south of Virgo, its head climbing westward beneath Leo.

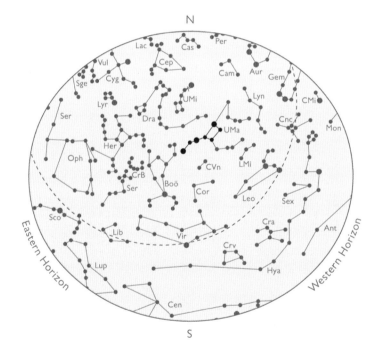

Timetable	Date	LMT	DST
	May 1	11pm	Midnight
	May 15	**10pm**	**11pm**
	June 1	9pm	10pm
	June 15	8pm	9pm

For key to constellation abbreviations, see page 27.

NORTHERN HEMISPHERE

The Milky Way, clearly in view, runs roughly north to south, east of centre across the sky. Straddling its dense starfields is the Summer Triangle (see page 47) of three brilliant stars: Vega in Lyra (Lyr), Deneb in Cygnus (Cyg) and, at the southeast apex, Altair in Aquila (Aql). Overlapping the triangle, Cygnus forms a figure known as the Northern Cross, with Deneb at the head. Far to the west, Virgo (Vir) with its bright star Spica is setting, but Arcturus in Boötes (Boö) still rides high. Toward the south a rare peek at Scorpius (Sco) with its bright star Antares can just be achieved on the horizon.

Timetable	Date	LMT	DST
	July 1	11pm	Midnight
	July 15	**10pm**	**11pm**
	August 1	9pm	10pm
	August 15	8pm	9pm

For key to constellation abbreviations, see page 27.

THE NIGHT SKY

NORTHERN HEMISPHERE

The Summer Triangle (see page 31) still dominates the sky but now appears toward the west; in full view toward the east flies the winged horse Pegasus (Peg). Touching the Square of Pegasus at its northeast corner is Andromeda (And). Cassiopeia (Cas), forming a flattened W-shape, swings in the circumpolar region directly north of Andromeda. North of Pegasus lies Cepheus (Cep). δ Cep (eastern arm) was the first-discovered Cepheid variable, varying a whole magnitude range in little more than 5 days. South of the Square, over the second fish of Pisces (Psc), and Aquarius (Aqr), is Fomalhaut in Piscis Austrinus (PsA).

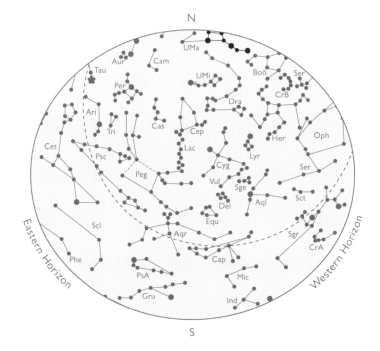

Timetable	Date	LMT	DST
	September 1	11pm	Midnight
	September 15	**10pm**	**11pm**
	October 1	9pm	10pm
	October 15	8pm	9pm

For key to constellation abbreviations, see page 27.

NORTHERN HEMISPHERE

The giant Orion (Ori), after a summer's absence, swings back boldly over the eastern horizon; to his west the river Eridanus (Eri) begins its winding flow. Centre-stage, along the ecliptic from Taurus (Tau) with its bright red eye Aldebaran, lies Aries (Ari). To the north, high above Aries' principal stars Hamal and Sheratan, shines Perseus (Per). By his western hand is the Double Cluster, excellent through binoculars, but appearing to the naked eye as a denser patch in the Milky Way. To the northwest, beyond the chained maiden Andromeda (And), is Cassiopeia (Cas), who continues her revolution around the pole.

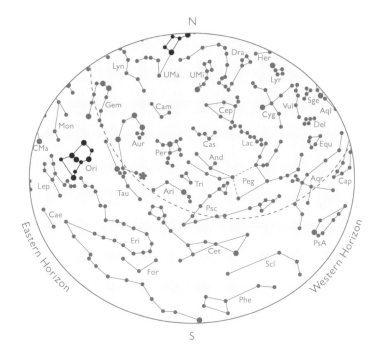

Timetable	Date	LMT	DST	For key to constellation abbreviations, see page 27.
	November 1	11pm	Midnight	
	November 15	**10pm**	**11pm**	
	December 1	9pm	10pm	
	December 15	8pm	9pm	

THE NIGHT SKY

SOUTHERN HEMISPHERE

Orion (Ori), head down, is unmistakable in the centre above the northern horizon. The three stars of his belt point northwest to Aldebaran in Taurus (Tau) and southeast to Sirius in Canis Major (CMa). An equilateral triangle of brilliant stars bridges the Milky Way: from the southern apex at Sirius, go to Betelgeuse at Orion's shoulder; complete the triangle eastward with Procyon in Canis Minor (CMi). To the south, the triangle points toward Canopus in Carina (Car). South of Carina lies the brilliant Southern Cross, Crux (Cru). A line through the vertical axis of the cross will point to the South Celestial Pole (see also page 162).

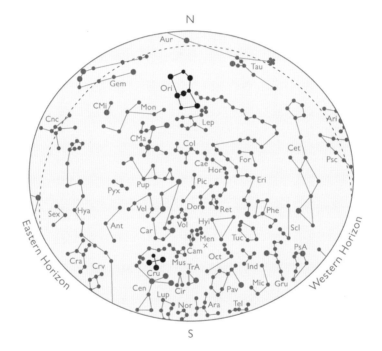

Timetable	Date	LMT	DST
	January 1	11pm	Midnight
	January 15	**10pm**	**11pm**
	February 1	9pm	10pm
	February 15	8pm	9pm

For key to constellation abbreviations, see page 27.

SOUTHERN HEMISPHERE

Orion (Ori) sets, followed by his dogs, Canis Major (CMa) and Canis Minor (CMi); their stars Sirius (the brightest in the sky) and Procyon dominate the western heavens. Leo has now come into the centre above the Northern horizon; the sickle that marks his head hangs from Regulus. Spica in Virgo (Vir) dominates the view east of Leo. Southwest, across the Milky Way, lies the second-brightest star in the sky, Canopus in Carina (Car). To the east of Carina lies the gentle Centaurus (Cen), the two bright stars in its feet making it unmistakable. Just rising over the eastern horizon is Scorpius (Sco).

Timetable

Date	LMT	DST
March 1	11pm	Midnight
March 15	**10pm**	**11pm**
April 1	9pm	10pm
April 15	8pm	9pm

For key to constellation abbreviations, see page 27.

SOUTHERN HEMISPHERE

Northward, beneath and to the east of brilliant Arcturus in Boötes (Boö), the beautiful Corona Borealis (CrB) sparkles on the horizon. Farther to the east, Hercules (Her) performs his deeds of valour. Above Boötes, a little to the west, Spica marks Virgo (Vir). High to the south lies the tiny form of the Southern Cross, Crux (Cru). Its horizontal beam points a short distance eastward along the Milky Way to Rigil Kentaurus in Centaurus (Cen), while its vertical axis points toward the South Celestial Pole. Beneath the feet of the Centaur lies Triangulum Australe (TrA). Scorpius (Sco) and Sagittarius (Sgr) rise into full view in the east.

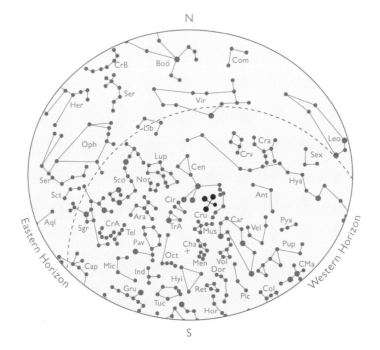

Timetable	Date	LMT	DST
	May 1	11pm	Midnight
	May 15	**10pm**	**11pm**
	June 1	9pm	10pm
	June 15	8pm	9pm

For key to constellation abbreviations, see page 27.

Ophiuchus (Oph), grappling with Serpens (Ser), is central above the northern horizon and Hercules (Her); the serpent's head can be seen just above the crescent Corona Borealis (CrB). Far up in the sky, the Milky Way groupings of Scorpius (Sco) and Sagittarius (Sgr) make fine viewing on the ecliptic. To the south the bright star Peacock can be seen in Pavo (Pav). Far south, east of the Milky Way, Achernar at the foot of Eridanus (Eri) becomes a prominent feature of the season's skies. Aquila (Aql) with its bright star Altair flies backward over the eastern horizon, as Virgo (Vir) begins to set in the west.

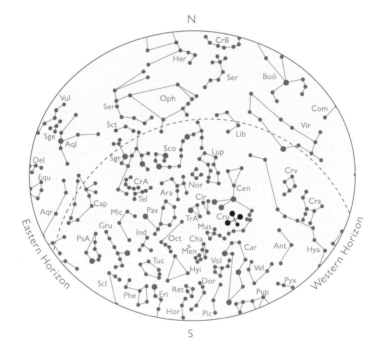

Timetable	Date	LMT	DST
	July 1	11pm	Midnight
	July 15	**10pm**	**11pm**
	August 1	9pm	10pm
	August 15	8pm	9pm

For key to constellation abbreviations, see page 27.

SOUTHERN HEMISPHERE

The winged horse Pegasus (Peg) races along the horizon from east to west, his famous Square showing clearly, toward the north. The sky overhead is dominated by the bright star Fomalhaut in Piscis Austrinus (PsA), while farther south, beyond Grus (Gru) and Phoenix (Phe), Achernar in Eridanus (Eri) is a brilliant sight. The Milky Way toward the west is intriguing, with Aquila (Aql) to the north, and Sagittarius (Sgr) and Corona Australis (CrA) to the south. Just west of Sagittarius is the hook of Scorpius (Sco). Northwest of Sagittarius, Ophiuchus (Oph) can be seen grappling with Serpens (Ser).

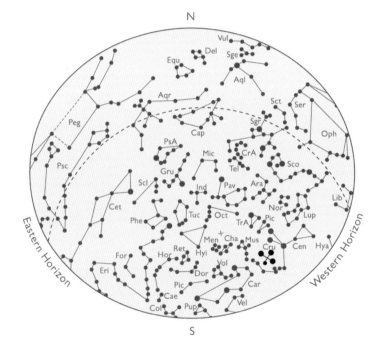

Timetable	Date	LMT	DST
	September 1	11pm	Midnight
	September 15	**10pm**	**11pm**
	October 1	9pm	10pm
	October 15	8pm	9pm

For key to constellation abbreviations, see page 27.

SOUTHERN HEMISPHERE

At the onset of Southern summer, Orion (Ori) pulls back into view, rising on the eastern horizon. From Rigel at his foot begins the faint outline of the long river Eridanus (Eri), winding toward Achernar, near the middle of our field and high in the south. Dropping northwest we come to Fomalhaut in Piscis Austrinus (PsA).

Toward the northern horizon note the bright stars of Aries (Ari). Southeast along the ecliptic from Aries, the beautiful cluster of the Pleiades in Taurus (Tau) can be seen over the horizon, the bright Aldebaran to their southeast. On the southern horizon lies Crux (Cru), pointing the way to the South Celestial Pole.

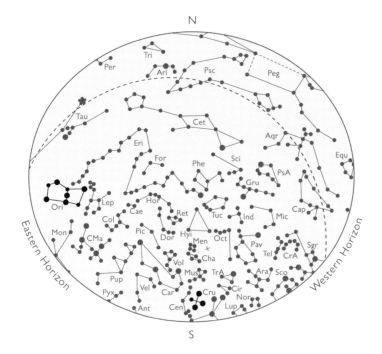

Timetable	Date	LMT	DST
	November 1	11pm	Midnight
	November 15	**10pm**	**11pm**
	December 1	9pm	10pm
	December 15	8pm	9pm

For key to constellation abbreviations, see page 27.

The Northern-hemisphere sky as shown by the mapmaker Carel Allard, *c.*1700. The constellations here are largely those noted by Ptolemy in the 2nd century CE, with the notable exceptions of Coma Berenices, Fluvius Jordanus and Tigris Fluvius.

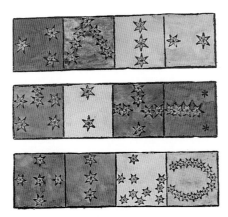

The 12 constellations of the zodiac from a woodcut, dated 1515. They are shown in highly stylized forms that appear to be unconnected with the shapes of the constellations that they represent. Each zodiacal constellation lies on the ecliptic.

THE MAJOR CONSTELLATIONS

" Some man of yore / ... thought he good to make the stellar groups, / That by each other lying orderly, they might display their forms. / And thus the stars / at once took names and rise familiar now. "

From an 18th-century translation of Aratus (3rd century BCE)

In this section, the first-rank and well-known constellations of both the Northern and Southern hemispheres are illustrated and explained both scientifically and mythologically. Of the 40 constellations treated here, only two (Crux and Canes Venatici; pages 85–6 and pages 56–7) were derived later than the formative classical period of Greco-Roman astrology and astronomy, two millennia ago; and one further figure, Carina (see pages 66–9), results from the modern division of a larger ancient constellation, Argo Navis. The naming and lore of individual stars often reflect the perceptions of later Persian and Arabic cosmographers, combining with the tradition handed down from Greece. To enter into the imagery of these figures is to pass through layers of history and culture toward the contemplation of ancient skies, a marvellous mythological realm.

ANDROMEDA

AND — ANDROMEDAE / THE ETHIOPIAN PRINCESS

The chained figure of Andromeda is fully visible for all latitudes as far south as 37°. She lies due west of the constellation that represents her saviour, Perseus, but is most easily located from the distinctive W of Cassiopeia, which lies immediately north. The head of the apparently falling Andromeda overlaps Pegasus at the horse's midriff, and the bright star here, Alpheratz, shares the northeastern corner of the "Square of Pegasus" (see pages 108–9). The constellation's midnight culmination is in the second week of October.

MAJOR STARS

α – Alpheratz or Sirrah, 2.06, blue-white
Both its common names are derived from the same Arabic phrase, *Al Surrat al Faras*, "the navel of the horse", as this star was at one time considered part of Pegasus (δ Peg). However, Arabic astrology derived from Ptolemy the title *Al Ras al Mar'ah as Musalsalah*, meaning "the head of the chained woman".

β – Mirach, 2.06, red
The name is derived from the Arabic for "girdle".

γ – Almach or Alamak, 2.26, orange
This star is named after a small animal similar to a badger, possibly derived from early Arabic or Persian sources. It marks the left foot of the girl.

M31 – The Andromeda Galaxy
M31 is a spiral galaxy similar in construction to our own. It appears as a fuzzy elliptical patch, and at a distance of 2.4 million light years it is the farthest object visible to the naked eye.

STARLORE

The background to the terrible events that result in Andromeda's being chained naked to rocks on the coast near Joppa (an ancient maritime city of Palestine) as an offering to the sea-monster Cetus is told fully under Cassiopeia (see pages 70–1).

As Andromeda lay helpless, Perseus, mortal son of Zeus, flew by on his return from his mission to slay the Gorgon Medusa. Some say that he was wearing the winged sandals given to him by Athene, the goddess of intellect and heroes. In another version, better suited to this group of constellation figures, he rode the winged horse Pegasus.

Approaching the rocks, Perseus was entranced by Andromeda's virgin beauty, and he offered to fight the sea-monster in return for her hand in marriage. Cepheus, her father, agreed to this transaction. Confusing Cetus with his shadow on the sea, Perseus dealt

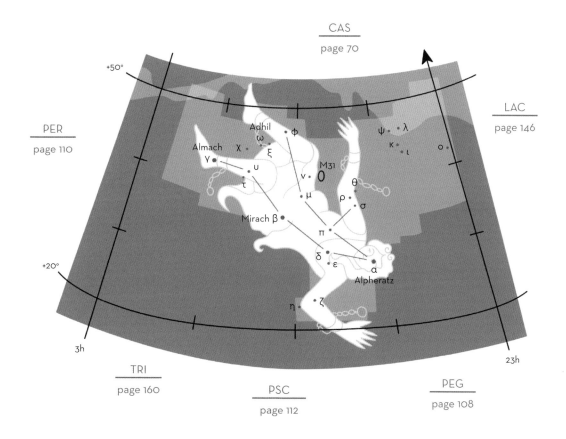

CAS
page 70

LAC
page 146

PER
page 110

TRI
page 160

PSC
page 112

PEG
page 108

the monster a fatal blow, and subsequently rescued Andromeda. (The constellations involved in the Andromeda story can be seen in their relative positions in the sky on Signpost Map 1, page 73.)

Behind the frame of Greek myth there is a darker and more puzzling root to this constellation. We see a hint of this in Andromeda's name, which means "ruler of men": as the Latin poet Manilius (1st century CE) remarks, "the vanquisher of Medusa was vanquished at the sight of Andromeda". So

perhaps she is not such a passive and innocent figure, but is closer to the goddess Aphrodite as a representative of female desire. This is borne out by the Mesopotamian roots of the Andromeda legend. In early times this constellation was dedicated to the ancient Egyptian goddess of love and war, Astarte (known as Ishtar to the Babylonians). Astarte, iconographically shown as a sexually voracious sea-goddess, was worshipped in various temples along the ancient shores of Palestine, the very shores of Andromeda's attempted sacrifice.

AQUARIUS

AQR – AQUARII / THE WATER-CARRIER

Aquarius, the 11th zodiacal constellation, is comparatively difficult to trace, with no star brighter than third magnitude. It was anciently depicted as a figure pouring water from a jug into the *Fluvius Aquarii*, the "River of Aquarius", which curves around beneath the water-carrier toward the bright star Fomalhaut in Piscis Austrinus. Fomalhaut remains a useful locator for Aquarius, and around 30° northwest of this star we are able to recognize the distinctive small cluster marking the water-jug. From Pegasus the jug and the head of the man can also be located, lying to the south, up against the head of the horse. The constellation culminates at midnight toward the end of August and early September.

MAJOR STARS

α – Sadalmelik, 3.0, yellow
This star marks the right shoulder of the figure, close by the jug. The name derives from the Arabic for "lucky stars of the king".

β – Sadalsuud, 2.9, yellow
The left shoulder is marked by this star, whose name means "luckiest of the lucky".

NGC 7293 – The Helix Nebula
At 300 light years, this is the nearest "planetary nebula" (see page 25) to our Sun. It covers a space about half the apparent size of the Full Moon, and is best seen through binoculars.

STARLORE

This constellation has carried a consistent mythical theme through various cultural transformations. The Babylonians, in the 2nd millennium BCE, showed the jar as an overflowing urn, and associated Aquarius with their 11th month (equivalent to our January–February), called the "curse of rain". The Egyptians saw the figure as a representation of Hapi, god of the Nile, who distributed the waters of life in heaven and on earth; his urn symbolized good fortune. This analogy fits the fortune associated with several of the stars by the jug and the waterman's head.

Late European representations of the water-carrier often show him as a bearded man, mature in years. However, the classical treatment is quite different. For the Latin poet Manilius (1st century CE), he is "the youth who pours, whom once [the Eagle] carried off from earth". This refers to the Greek myth of the boy Ganymede, whose name means "rejoicing in virility". The son of King Tros of Troy, Ganymede

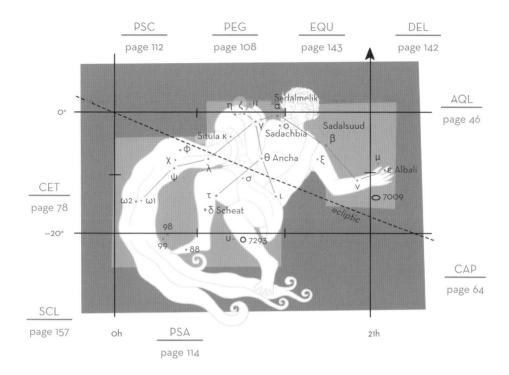

PSC
page 112

PEG
page 108

EQU
page 143

DEL
page 142

AQL
page 46

CET
page 78

CAP
page 64

SCL
page 157

PSA
page 114

Sadalmelik
α

η ζ ι π
Sadachbia
Sadalsuud
β

γ ο
Situla κ
θ Ancha
ξ
μ
φ
ε Albali
χ
λ
ν
ψ
σ
ο 7009
ω2 ω1
τ
ι
δ Scheat
98
υ ο 7293
99 88

ecliptic

0°

−20°

0h

21h

was the most beautiful youth alive, and was chosen by the gods to be for ever young as the bearer of the golden cup of divine nectar. In later versions, Zeus (in Roman myth, Jupiter), king of the gods, desired the boy. Disguised as the eagle of Aquila (see pages 46–7), Zeus is said to have carried Ganymede up to Olympus to become his personal cup-bearer.

Unsurprisingly, Zeus' abduction of Ganymede caused repercussions on Mount Olympus. The arrival of the youth as cup-bearer displaced Hebe, goddess of youth and a daughter of Hera, Zeus' queen. Hera was angered by the insult to Hebe, and by her shame at Zeus having fallen in love with a boy. This attitude enraged Zeus, who glorified Ganymede by setting him among the stars.

An illumination from the *Bedford Book of Hours*, which dates from c.1423, representing January–February by the zodiacal sign of Aquarius.

AQUILA

AQL – AQUILAE / THE EAGLE

Aquila is a small but beautiful constellation to the south of Cygnus. Culminating at midnight in July, the eagle appears to fly eastward across the Milky Way. Lying on the equator, it can be seen from all but the most extreme Northern and Southern latitudes. Its brilliant lucida, Altair, forms one apex of the Northern-hemisphere Summer Triangle (see opposite, below).

MAJOR STARS

α – Altair, 0.77, white
The 12th brightest star in the sky, Altair takes its name from the Arabic word for the eagle, and thus represents the whole constellation.

β – Alshain, 3.7, yellow
The name of this star is derived from the Persian name for the whole constellation.

γ – Tarazed, 2.7, yellow
Together, the α, β, and γ stars form the distinct close group known as the "Family of Aquila"; the line connecting them is just 5° of arc.

STARLORE

This constellation has Mesopotamian origins, being represented as an eagle in a stone relief dating from as early as c.1200 BCE. For the Greeks the eagle, like all creatures of the air, came under the dominion of the supreme god Zeus (in Roman myth, Jupiter). The eagle was the king of the birds, a privileged royal servant and a fighter, charged in particular with retrieving the thunderbolts hurled by the great sky-god.

One particularly relevant myth concerns the abduction and seduction by Zeus of a beautiful boy, Ganymede. Pictorially, Aquila was sometimes represented carrying the youth skyward in his claws. Ganymede is shown by the stars toward the south of the constellation, with his head at β Aql (see also Aquarius, pages 44–5).

Another myth reveals the ferocious nature of the eagle. In some accounts, Prometheus – whose name means "foresight" – was one of the last generation of Titans, descendants of the primal deities Uranus and Gaea. He is also said to be the creator and divine protector of human beings. Prometheus taught humankind the arts and sciences, which Zeus considered too great a gift to bestow on the inferior human race. Still, Prometheus persisted, supplying them with the gift of fire, which he took from the Sun and secretly smuggled to Earth in a hollow fennel stem.

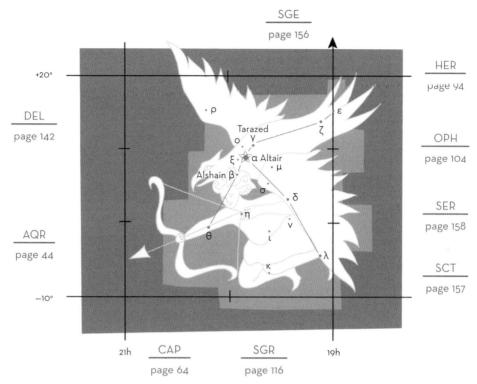

SGE
page 156

HER
page 94

DEL
page 142

OPH
page 104

SER
page 158

AQR
page 44

SCT
page 157

CAP
page 64

SGR
page 116

+20°

−10°

21h

19h

ρ

ε

Tarazed
ο γ

ζ

ξ α Altair
μ

Alshain β

σ

δ

η

ν

θ

ι

κ

λ

Zeus, enraged by this act, devised a terrible punishment for Prometheus. He was chained naked to a pillar in the Caucasus Mountains, and from dawn to dusk the eagle of Zeus tore through his flesh to his liver. But because Prometheus was immortal, his liver healed every night, only to be pecked out again when the eagle reappeared at dawn the following day. In this way his suffering was destined to go on for ever. However, many years later Zeus accepted an appeal from the hero Heracles (Hercules) to show Prometheus mercy: the wise centaur Chiron (see pages 74–5) agreed to relinquish his immortality in exchange for Prometheus' freedom. Once Zeus had relented, Heracles shot the eagle through the heart.

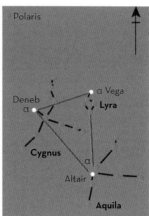

The Northern-hemisphere Summer Triangle is formed by the lucidae Altair in Aquila, Vega in Lyra and Deneb in Cygnus. The brightest is Vega with a magnitude of 0.

THE MAJOR CONSTELLATIONS

ARIES

ARI — ARIETIS / THE RAM

Aries is the first zodiacal constellation, but despite its mythological significance the figure, which lies to the west of Taurus, is not distinctive, apart from the close group of three stars, α, β and γ, marking the head of the ram. Hamal (α Ari) lies on the meridian (north-south) line which runs up to the North Pole from this star through Almach in Andromeda (γ And) and Segin in Cassiopeia (ε Cas). Extrapolating the same line southward to the equator brings us a few degrees to the west of Mira in Cetus (ο Cet).

MAJOR STARS

α — Hamal, 2.0, yellow
Named after the Arabic for "lamb", this star culminates at midnight around October 22.

β — Sheratan, 2.6, white
From the Arabic for "mark" or "sign", the name Sheratan was at one time used jointly for this star and Mesarthim (γ Ari). The stars were so called because they marked the March equinox point, 0° Aries, which fell close by here c.300–400 BCE.

STARLORE

Marking the March equinox point, Aries was held in high status in the formative period of Greek skylore. The Roman poet Manilius (1st century CE) declared it "the prince of all the Signs". The Assyrians of the Upper Tigris, who would sacrifice a ram in honour of the equinox, knew the constellation as "altar" and "sacrifice".

In Greek lore, Aries represents the legend of the Golden Fleece. According to the poet Apollonius of Rhodes (3rd century BCE), King Athamus of Boetia married Nephele. Athamus became disenchanted with his wife and remarried. His new wife, Ino, saw the children of the king's previous marriage, especially the boy Phrixus, as a threat to her own offspring. She therefore wickedly devised an ingenious scheme to bring about the child's death. She went secretly to the stores of corn seed kept for sowing in the spring, and scorched the precious grains. The resulting crop failure threatened the population with starvation. Athamus sent a messenger to the oracle at Delphi, but Ino had already bribed the emissary, who returned saying that the oracle required the young prince as a sacrifice before the corn would grow again. Phrixus was duly prepared for sacrifice, but hearing the desperate prayers of Nephele, Hermes, the messenger of the gods, intervened

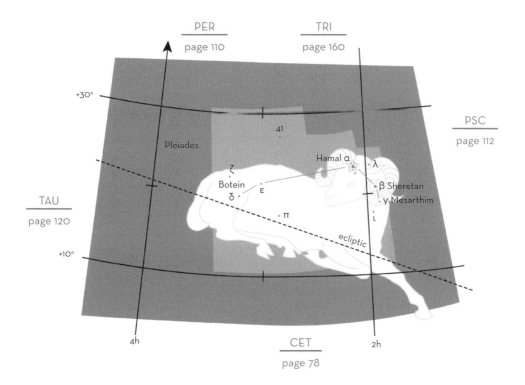

PER
page 110

TRI
page 160

PSC
page 112

TAU
page 120

CET
page 78

+30°

+10°

41

Pleiades

Hamal α · λ

Botein
δ · ε ζ

β Sheretan
γ Mesarthim
ι

π

ecliptic

4h

2h

and sent a wonderful ram with a fleece of gold to snatch the boy from the altar.

Phrixus had a sister, Helle, who was also rescued by the ram, but as the magical beast crossed the narrow straits that separate Europe and Asia, Helle fell to her death. From that time on, these straits were called the Hellespont ("sea of Helle") in her memory.

The ram brought Phrixus to the land of Colchis on the shores of the Black Sea. Here Phrixus, in gratitude for his salvation, sacrificed the ram to Zeus (Jupiter in Roman myth), and gave its Golden Fleece to King Aeëtes of Colchis. Aeëtes had the Fleece protected by a dragon in a sacred grove of the war-god Ares (Mars), just as in later astrology the zodiacal sign of the ram is placed under the dominion of the war-god. The Fleece remained in the grove until stolen by the hero Jason (see pages 66–8).

Aries, from an Italian manuscript illustration (9th–10th century CE). Aries is associated with the Greek god Ares (in Roman myth, Mars). Accordingly, in astrology Aries rules the planet Mars.

AURIGA

AUR – AURIGAE / THE CHARIOTEER

A striking constellation of the Northern-hemisphere winter skies, Auriga lies to the north of the horns of Taurus. The two constellations actually join at the northern horn of the bull, marked by Elnath (β Tau), for this star doubles up as the right foot of the charioteer. Once recognized, the distinctive curve of stars in Auriga is never forgotten; seen in the Northern hemisphere the curve runs upward in a clockwise arc from Elnath, through ν to β (the bright Menkalinan), and round to brilliant Capella (α, the figure's lucida). Below and to the south of Capella at the far end of the curve we come to the little cluster of stars marking the kids. Auriga's midnight culmination is in December.

MAJOR STARS

α – Capella, 0.08, yellow-white
The name means "she-goat". This is the sixth brightest star in the sky, and lies at a distance of 46 light years from Earth.

β – Menkalinan, 1.90, yellow
The name for this star derives from the Arabic for "charioteer's left shoulder".

M36, M37, M38
Star clusters worth investigating with binoculars. M37 lies at a distance of 4,400 light years.

STARLORE

Represented as a charioteer in Mesopotamia, Auriga from earliest times has been shown cradling a goat or kid goats – later said to be the goat Amaltheia, of Greek myth, who suckled Zeus (Jupiter). Some say the figure is Erichthonius, son of Mother Earth and Hephaestus. Ericthonius introduced the four-horse chariot into Athens.

In another interpretation Auriga represents the ill-fated charioteer Myrtilus. King Oenomaus, noted for his love of horses, could not bear the thought of his daughter Hippodameia ("horse-tamer") marrying. He devised a chariot race, in which he would race each suitor for her hand, the suitor forfeiting his life if the king won. Oenomaus' horses, swifter even than the North Wind, came from the god Ares (Mars) and were unbeatable, so that Oenomaus was able to slay each of his daughter's suitors.

When it came to the turn of Pelops, son of Hermes, the gods decided to intervene.

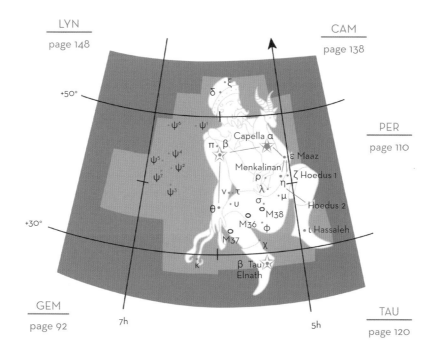

LYN
page 148

CAM
page 138

PER
page 110

GEM
page 92

TAU
page 120

+50°

+30°

7h

5h

δ · ξ

Capella α

π · β

ε Maaz

ψ⁶ · ψ¹

Menkalinan
ρ · ζ Hoedus 1
λ η

ψ⁵ · ψ⁴
· ψ²
ψ⁷ · μ

ψ³

ν · τ
σ · Hoedus 2
θ · υ

M38

M36 φ

M37

ι Hassaleh

κ

β Tau ☆
Elnath

χ

Poseidon (Neptune), an ancient god of horses as well as ruler of the seas, gave Pelops a gold chariot drawn by winged golden steeds. To further ensure his victory, and with the collusion of Hippodameia, Pelops plotted with Myrtilus, Oenomaus' charioteer, to replace the lynchpins from the axles of the king's chariot with copies made of wax. Pelops promised that if the king lost the race, then the crafty charioteer would be given half the kingdom and the privilege of the bridal night with Hippodameia. At the climax of the race the wheels flew off Oenomaus' chariot and he was dragged to his death, uttering a curse on Myrtilus as he died.

Celebrating their victory, Pelops, Hippodameia and Myrtilus went for a chariot drive. When they stopped the chariot to have a picnic, Myrtilus demanded part of his reward there and then, but Hippodameia resisted.

Pelops struck the lustful charioteer, took the reins and began the journey home. As they sped back Pelops gave Myrtilus a sudden kick, hurling him to his death. Hermes, who appreciated a cunning trick, honoured the charioteer by placing him in the stars.

Auriga, holding a kid in his left arm and the charioteer's whip in his right hand, from a star map of *c*.1660.

BOÖ – BOÖTIS / THE HERDSMAN

Boötes is a prominent constellation for the Northern-hemisphere spring and early summer skies, culminating at midnight around May 1. Lying to the northeast of Virgo, its outstanding star, Arcturus, is fourth brightest in the heavens. Boötes cannot be seen as a complete figure for much of the Southern hemisphere beyond the Tropics, but the location of Arcturus in the south of the constellation means that for middle-latitudes South this impressive star comes into full autumn view, below Virgo, but well above the Northern horizon.

MAJOR STARS

α – Arcturus, –0.04, rich golden yellow
Arcturus, which lies 36 light years distant, is believed to have roughly the same mass as our Sun but 27 times the diameter. In it we see prefigured the fate of the Sun, which is expected to swell to become a similar red giant some 5,000 million years from now. Arcturus was one of the most closely observed stars in antiquity, and is mentioned by the Greek poet Hesiod in the 8th century BCE. The name means "bear-keeper", referring to Boötes' perpetual pursuit of Ursa Major and Ursa Minor (the Greater and Lesser Bears) around the North Pole. The star had an ancient reputation as a harbinger of storms, but in later astrology was seen as a bringer of wealth and honour.

β – Nekkar, 3.5, yellow
The name derives from the Arabic for "ox-driver", a name also given to the whole constellation.

γ – Seginus, 3.0, white
This star is also known as Haris. The name Seginus was sometimes applied to the whole constellation, but its origin is obscure.

ε – Izar, 2.7, orange
"Loincloth" or "belt". This much-studied star is a double, having a blue, fifth-magnitude companion. The exquisite contrast, seen when the double is separated into its components (through a medium-powered telescope), has led to the alternative name *Pulcherrima*, "most beautiful".

η – Muphrid, 2.7, yellow-white
The name is taken from the Arabic *Al Mufrid al Ramih*, "solitary star of the lancer"; Boötes was often shown waving a sword or lance, but the location of this star in the leg is inexplicable.

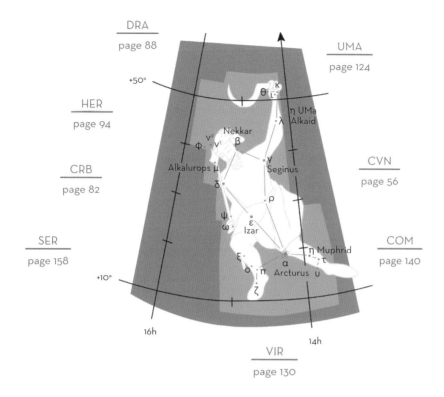

DRA
page 88

UMA
page 124

HER
page 94

CRB
page 82

CVN
page 56

SER
page 158

COM
page 140

VIR
page 130

STARLORE

One legend identifies Boötes with the Athenian Icarius, whose daughter was Erigone. The god Dionysius (in Roman myth, Bacchus) taught Icarius the secret of winemaking. Subsequently, Icarius gave wine to some peasants, who became drunk. Thinking that they had been poisoned, the peasants slaughtered Icarius and buried his body.

Aided by her father's dog Maera, Erigone searched for Icarius' grave. When she found it, she was so grief-stricken that she hanged herself. Zeus (or some say Dionysius) placed her in the heavens as Virgo. Icarius became Boötes and Maera became, alternatively, Procyon in Canis Minor (see page 62) or one of the dogs of neighbouring Canes Venatici.

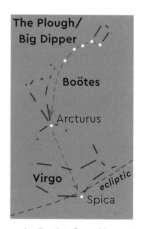

Arcturus (α Boö) is found by extending the curve of the handle of the Plough or Big Dipper of Ursa Major toward the southeast. Continue the curve to find Spica (α Vir) in Virgo, lying on the ecliptic.

THE MAJOR CONSTELLATIONS

CANCER

CNC – CANCRI / THE CRAB

Cancer is the least conspicuous of the 12 zodiacal constellations – not one of its stars is brighter than the fourth magnitude. It covers a modest corner of sky between the dominant figures of Gemini to the west and Leo to the east, and its most interesting visual feature is the Beehive Cluster M44, known in classical times as Praesepe (meaning "the manger"). Cancer's midnight culmination occurs at the end of January or at the beginning of February.

MAJOR STARS

α – Acubens, 4.3, white
The name means "claw".

α and δ – Asellus Borealis and Asellus Australis, 4.7 and 4.2, both pale yellow
The northern and southern asses. These stars bracket the Beehive Cluster.

M44 – The Beehive Cluster or Praesepe
A group of some 50 stars of the sixth magnitude and fainter, 520 light years away. They are visible to the naked eye as a cloudy patch three times the diameter of the Moon.

STARLORE

The fixed stars of Cancer once marked the position of the Sun at the June solstice. For the Mesopotamians this key station marked the gateway for the descent of souls into incarnation. This analogy is consistent with Egyptian tradition, where Cancer was the dawn Sun-god Khephri, a heavenly personification of the humble scarab, or dung beetle, symbolizing fertility, life and rebirth. To the Greeks Cancer was the crab that tried to nip the toes of Heracles (see pages 94–5) as he battled with the monstrous Hydra.

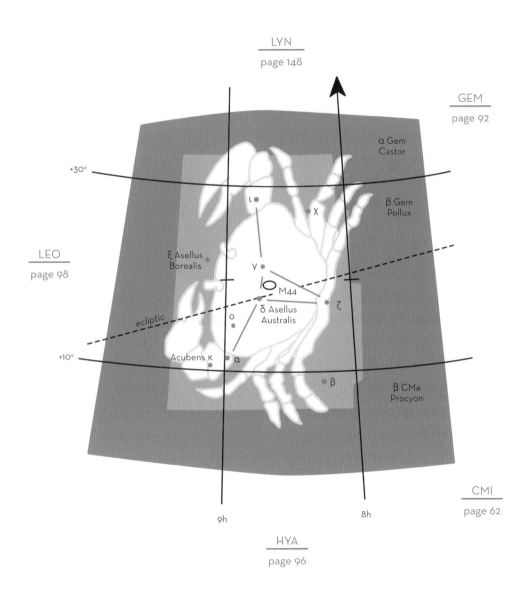

LYN
page 148

GEM
page 92

α Gem
Castor

β Gem
Pollux

+30°

χ

ι

LEO
page 98

ξ Asellus
Borealis

γ

M44

ζ

ecliptic

δ Asellus
Australis

ο

+10°

Acubens κ   α

β

β CMa
Procyon

CMI
page 62

9h

8h

HYA
page 96

THE MAJOR CONSTELLATIONS

CANES VENATICI

CVN – CANUM VENATICORUM / THE HUNTING DOGS

The Hunting Dogs lie to the west of their master Boötes, and beneath the tail of Ursa Major. Two stars of note, Cor Caroli and Chara, mark the collar and head of the southern hound.

The northern hound is hardly traceable in the faint stars of its area of sky, but it does cover the Whirlpool Galaxy, M51. Canes Venatici culminates at midnight in early April.

MAJOR STARS

α – Cor Caroli, 2.9, white
"The heart of Charles", after the executed Charles I of England. It is said to have shone out on May 29, 1660, when Charles II returned to London, heralding the restoration of the monarchy.

β – Chara, 4.3, yellow
The name seems to have derived from the Latin for "edible root".

M51 – The Whirlpool Galaxy
This is an eighth-magnitude spiral galaxy around 15 million light years distant, and it makes an intriguing object for a binocular search under good viewing conditions.

STARLORE

The 17th-century groupings of Polish astronomer Johannes Hoevelke ("Hevelius") have generally endured the test of time, while those of other modern cartographers have not. Following Hevelius, Canes Venatici is shown on all modern representations as two hounds at the heels of Ursa Major, held on a leash by Boötes. A leading modern interpreter of stellar myth, Julius Staal (1917–1986), locates these stars as the dogs that guided the daughter (Virgo) of Icarius (Boötes) to their dead master's body (see page 53), in preference to the more ancient – but less astronomically fitting – Canis Minor.

UMA

page 124

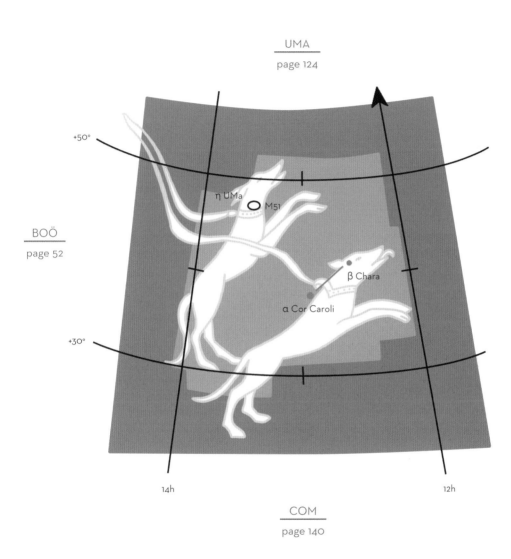

BOÖ

page 52

+50°

η UMa

○ M51

β Chara

α Cor Caroli

+30°

14h

12h

COM

page 140

CANIS MAJOR

CMA – CANIS MAJORIS / THE GREATER DOG

The constellations of the two dogs (Canis Major and Canis Minor) of Orion dominate their regions of sky by virtue of the great stars Sirius (α CMa) and Procyon (α CMi). For both, the constellation myths are often an extension of the stories of these individual stars. Canis Major lies south of the equator, making it a striking sight from the Tropics and the Southern hemisphere. Upper- and middle-latitude observers in the North often lose the full visual impact of Sirius when it is seen low on the horizon (see page 21).

MAJOR STARS

α – Sirius, 1.46, brilliant white
The name means "scorching". The brightest star in the heavens, Sirius can be outshone only by a planet. It lies 8.7 light years distant, making it one of the Sun's closest neighbours. It marks the head or jaw of the dog. There is a controversial claim that the Dogon peoples of Mali in West Africa have traditionally given Sirius a companion-star, Po. This they termed the "heaviest star", and calculated their ritual time-periods on the basis of its 50-year, elliptical orbit. However, not until 1862 was it scientifically proven that Sirius was actually a binary (double star), its tiny companion Sirius β (magnitude 8.5) orbiting every 50 years. How the Dogon made this discovery hundreds of years earlier remains a mystery.

β – Mirzam, 2.0, blue-white
The name means "announcer", possibly because the star rises slightly ahead of Sirius.

δ – Wezen, 1.8, yellow
The name means "weight" in Arabic; but the reason for this is not clear.

ε – Adhara, 1.5, blue
The name comes from the Arabic for "virgins", and the legend of the two sisters associated with Canis Minor (see page 62) may be equally relevant here.

STARLORE

The dog symbolism of Canis Major and its lucida Sirius goes back to at least the 3rd millennium BCE. During this epoch, Sirius, also known as Sothis, was the marker star for the Egyptian Sothic calendar. Its heliacal rising (its first brief appearance just before sunrise, after a period of several months unseen) occurred in mid-July and coincided with the annual rising of the Nile:

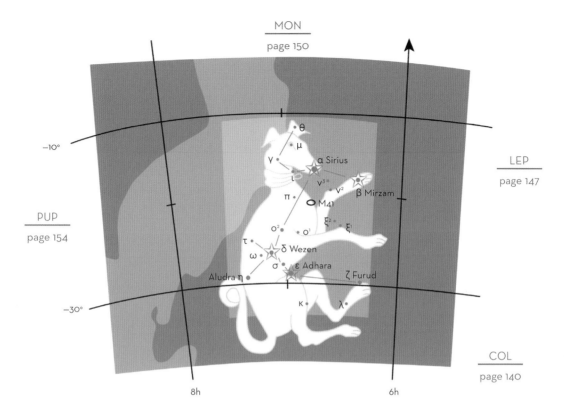

MON
page 150

LEP
page 147

PUP
page 154

COL
page 140

the flooding of the river basin was paramount for the fertility of the land, and was therefore the basis of the life and wealth of Egypt.

Over the long span of Egyptian history Sirius/Sothis took on several layers of interpretation, including an identification with the goddess Isis, sister and consort of the great god Osiris, linked with Orion. Eventually, as the cult of Isis broadened to merge with that of the cow-goddess Hathor, Sirius became the star of Isis-Hathor, depicted as having the horns of a cow. However, it is the dog symbol that has the earliest origins. Sirius was at one time identified with the jackal-headed god Anubis, who like the Greek Hermes was a guide of the dead. Anubis invented the art of

embalming and was the lord of funeral rites; he also weighed dead souls on the scales of justice to determine their fate in the afterlife.

Egyptian tradition further associated Sirius with the "Dog Days", an identification that is believed to be the origin of this star's naming as the "Dog Star". Dog Days originally referred to the 40-day period at the start of each Sothic year, when the summer was at its hottest. Classical authors often identified the power of Sirius with that of the Sun, and the star was sometimes represented with a corona of rays. The name Sirius is derived from the Greek name, Serios, meaning "scorching", and the star was held to provoke a deadly fever, exemplified by rabies in mad dogs.

The Greeks adopted the earlier lore concerning Sirius, but additionally wove the whole constellation into the fabric of their own mythology. Both Canis Major and Canis Minor were seen as belonging to the hunter Orion (see pages 106–7); and in Mesopotamian stellar mythology we find the same image of a dog at the heels of a giant man, watching as if to pounce on the hare Lepus (see page 147) at Orion's feet.

Several authors, including the Roman poet Ovid (43 BCE–17 CE), saw one or other of the two dogs as Maera, the faithful dog of Icarius (represented by Boötes, pages 52–3): Maera's name means "shining". However, the modern constellation Canes Venatici is also a plausible contender for this role (see page 52).

Another tradition of imagery has pictured Canis Major as the terrible Cerberus, the three-headed hound who in Greek myth guards the entrance to the underworld Hades. As so often with myth, an apparently disconnected jumble of images yields hidden threads of association. Cerberus guards Hades, which is the realm of the dead, which reminds us that Anubis (Sirius) was a guide to the dead and could enter the forbidden realm. The story of Maera falls into place as well, because it was this dog who led Erigone to the buried corpse of her father Icarius — once again, guarding or serving the dead is an underlying theme.

In an interesting cross-cultural parallel, for the Chinese Sirius was T'ien-lang, the celestial jackal. The southern stars of Canis Major represented the bow and arrow that were used to kill T'ien-lang, after he had ravaged the crops of the Chinese king.

Sirius can be found easily from the belt of Orion. Locate the belt in the sky and extend a straight line through its stars toward the southeast. This line points directly to Sirius.

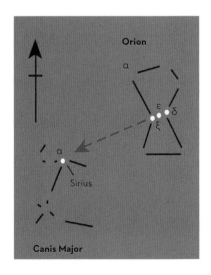

The jackal-headed god Anubis shown tending the dead in an Egyptian tomb painting. The star Sirius (α CMa) was thought to represent Anubis. In some accounts Anubis is the son of the Egyptian god Osiris, who was identified in the sky with the constellation Orion, the supposed celestial owner of the Greater and Lesser Dogs.

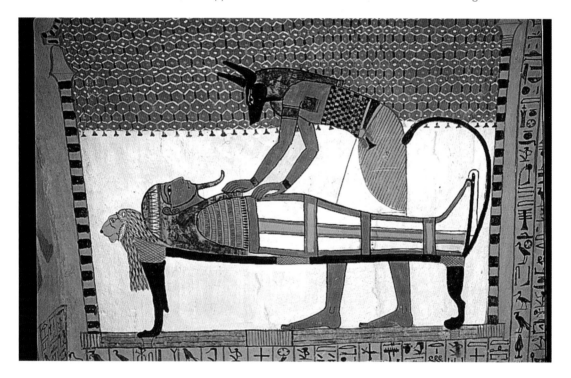

CANIS MINOR

CMI – CANIS MINORIS / THE LESSER DOG

Canis Minor is a small constellation whose myth is often identified solely with its lucida, Procyon. It lies due south of the Gemini twins on the other side of the Milky Way to Canis Major. If you have any trouble locating these stars, extend a line eastward from Bellatrix (γ Ori) in the left shoulder of Orion to Betelgeuse (α Ori) in the giant's right shoulder, and this will run directly to Procyon. Furthermore, Procyon, Sirius (α CMa) in Canis Major and Betelgeuse form an equilateral triangle of first-magnitude stars.

MAJOR STARS

α – Procyon, 0.4, yellow-white
Lying 11.4 light years from Earth, this is one of the nearest stars to our Sun. It is the eighth brightest star in the sky. A double star, its companion is a white dwarf of very dim magnitude 10.3, with an orbital period of 41 years. The name Procyon comes from the earliest Greek records and means "before the dog", which suggests that the star was seen, like β CMa (Mirzam), as announcing the rising of Sirius.

β – Gomeisa, 2.9, blue-white
The name is derived from an Arabic alternative name for the whole constellation, "watery-eyed" or "weeping one" (see below).

STARLORE

The Mesopotamians saw Canis Minor as a water-dog, which may be the origin of the occasional Arabic designation of Procyon as *Al Ghumaisa*, "watery-eyed"; but this equally fits the Arabic legend that Canis Major and Minor are two sisters, one of whom (represented by Canis Major) eloped, leaving her desolate sibling behind.

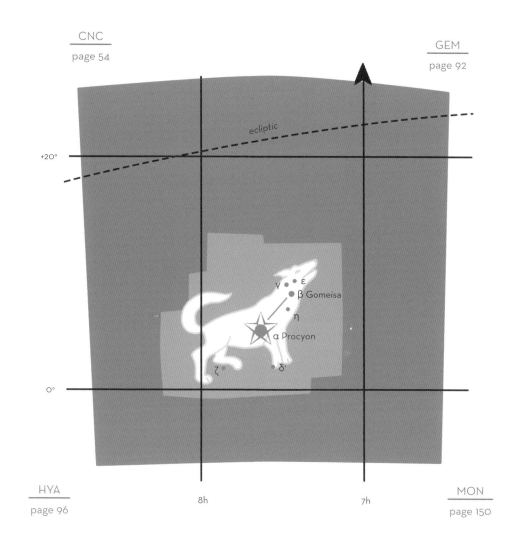

CNC
page 54

GEM
page 92

ecliptic

+20°

γ ε
β Gomeisa
η
α Procyon
ζ δ

0°

HYA
page 96

8h

7h

MON
page 150

THE MAJOR CONSTELLATIONS

CAPRICORNUS

CAP – CAPRICORNI / THE SEA-GOAT

The tenth and smallest zodiacal constellation, Capricornus is traced out by third- and fourth-magnitude stars, east of Sagittarius. Its midnight culmination is in early August, but the combination of light skies and a location south of the equator renders it unimpressive on a summer evening from middle and upper latitudes of the Northern hemisphere. It can be found by a line from Vega (α Lyr), across the Milky Way, through Altair (α Aql) to Algedi and Dabih, the α and β stars of the horns of the goat.

MAJOR STARS

α – Algedi or Giedi, 3.6, yellowish
Both names mean either "goat" or "ibex". Algedi is actually a pair of stars, apparently close but unrelated. The name *Dabih* (see below) has also been given to this star.

β – Dabih, 3.1, golden-yellow
The name is derived from the Arabic *Al Sa'd al Dhabih*, the "lucky one of the slaughterers", referring to the Arabic tradition in which a goat was sacrificed when the Sun first entered the star-fields of Capricorn.

γ – Nashira, 3.8
The name derives from the Arabic for "bringer of good tidings".

δ – Deneb Algedi, 2.9
The "tail of the goat", Capricorn's lucida. 5° to its east is the point at which, in 1846, the French astronomer Le Verrier calculated the existence of the planet Neptune – a delightful mirroring of the mythological connection between Capricorn, Neptune and the sea.

STARLORE

For the Mesopotamians Capricornus marked the point in the year when the Sun was at its farthest point south of the equator – the December solstice. The imagery of Capricorn as a goat-fish may have Assyro-Babylonian origins in Oannes, god of wisdom, who was half-fish, half-man. This figure reappeared at long intervals in the Persian Gulf, disguised as a mermaid, and taught humankind the arts and sciences.

Among the Latin poets Capricornus was known as *Neptuni proles*, "Neptune's offspring" (the Roman god Neptune, or Poseidon in Greek myth, ruled the sea). In Indian starlore the constellation was a crocodile, or a curious goat-headed hippopotamus.

Aside from the goat-fish theme, Capricornus is associated with the Greek Pan (Priapus in Asia Minor), noted for his lustful behaviour

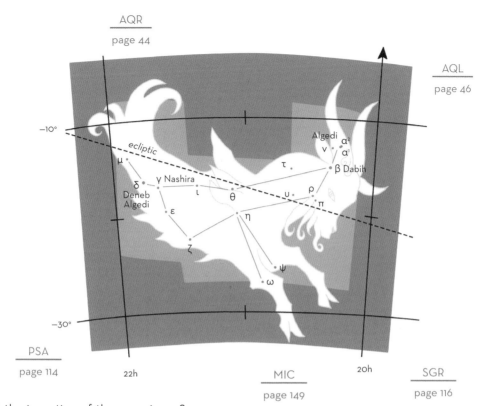

AQR
page 44

AQL
page 46

PSA
page 114

SGR
page 116

MIC
page 149

Algedi
ν · α²
· α¹

β Dabih

τ ·

γ Nashira

μ ·

δ ·
Deneb
Algedi

ι

θ

υ ·
ρ ·
π

ε

η

ζ

ψ

ω

ecliptic

−10°

−30°

22h

20h

and the invention of the pan pipes. Some say that he was a satyr, a man with goat's legs, cloven hooves and small horns. He became honoured when the sea-monster Typhon was sent by the Titan goddess Rhea to destroy the Olympian gods. As the monster approached, Pan plunged into a river and tried to turn himself into a fish to escape. However, he managed only half his transformation. By the time that he had struggled back onto land, Typhon had dismembered the supreme god Zeus (Jupiter). To frighten the monster Pan let out a shriek long enough to enable nimble Hermes (Mercury) to retrieve Zeus' scattered limbs. Together Pan and Hermes carefully restored the god, who then rewarded Pan, now a satyr-fish, by placing him in the constellations.

Capricornus, with the head of a goat and the tail of a fish, from a medieval English manuscript.

THE MAJOR CONSTELLATIONS

CARINA

CAR – CARINAE / THE KEEL OF THE SHIP

This constellation was originally part of a much larger Southern-hemisphere figure, Argo Navis (the Ship). In 1763, the French cartographer de Lacaille divided the larger constellation into its constituent sections, Carina, Puppis and Vela. Carina lies partly on the Milky Way, well to the south of Sirius (α CMa) and Procyon (α CMi), and is obscured from middle-Northern latitudes upward. The constellation culminates at midnight in the Southern midsummer months: its brilliant lucida, Canopus, culminates at midnight around December 28.

MAJOR STARS

α – Canopus, −0.7, white

The second brightest star in the sky, this supergiant lies at 205 light years distance from Earth. It is a major navigational star, and is used by NASA as a marker for setting space-flight coordinates. Its name may originate from the Coptic or Egyptian *Kahi Nub*, which means "golden earth". In Greek legend King Menelaus pillaged Troy in 1183 BCE. The helmsman of his ship, Canopus, is represented in the sky by the star that bears his name – a fitting link to Canopus' modern association with navigation.

β – Miaplacidus, 1.7, blue-white

This star is 55 light years away. The origin of the name is uncertain.

STARLORE

From ancient times Argo Navis was associated with the archetype of a great ship, which crosses the waters of the Deluge, as in the Biblical tale of Noah's Ark. Similarly, the Babylonian Epic of Creation relates how the gods decided to destroy the Earth with a flood. The god Ea took pity on humanity, and secretly warned a mortal by the name of Uto-Napishtim of the forthcoming disaster. The man set about building a huge boat 120 cubits high to carry his family, precious possessions and sundry animals and birds. After the flood subsided, Uto-Napishtim and his passengers were the only survivors.

For the Greeks Argo was the ship of the hero Jason and his crew, the Argonauts. The story begins with the boy Phrixus fleeing from his murderous stepmother to Colchis on the back of the Golden Ram (see Aries, pages 48–9). The ram became a sacrifice to the great god Zeus (in Roman myth, Jupiter), and its magical fleece was kept in a grove sacred to the god Ares (Mars), guarded by a fierce dragon that never slept. Phrixus remained in Colchis and eventually married the king's

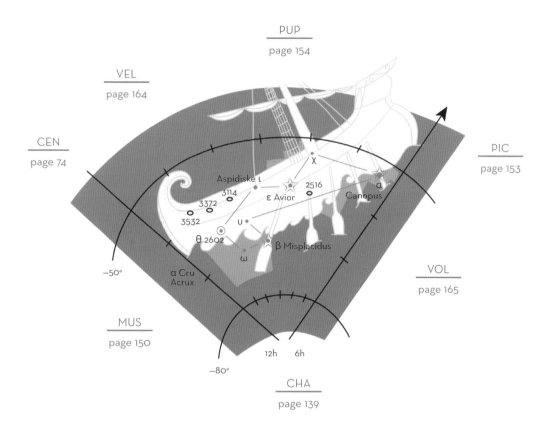

PUP
page 154

VEL
page 164

CEN
page 74

PIC
page 153

VOL
page 165

MUS
page 150

CHA
page 139

Aspidiske ι
3114
ε Avior
2516
Canopus
α
3372
3532
υ
θ 2602
β Misplacidus
ω
α Cru
Acrux

−50°

−80°

12h   6h

χ

daughter, thereby becoming heir to the throne, but he remained in exile from his native land. Even after death his soul was restless, and an oracle warned that this would bring a curse of barrenness to his homeland in the succeeding generation.

Later, King Aeson of Iolcus in Thessaly (Phrixus' homeland) was usurped by his half-brother Pelias. Jason, the baby son of the dethroned king, barely escaped with his life and was taken by his mother to be brought up by the wise centaur Chiron (see Centaurus, pages 74–5). As a youth Jason asserted his claim to the throne of Iolcus. Pelias agreed, on condition that Jason should lift the curse of Phrixus by bringing the Golden Fleece, and therefore Phrixus' soul, back to Thessaly.

Jason took up the challenge and 50 of the greatest heroes came forward to serve as his crew. A ship with 50 pairs of oars, named the *Argo*, was constructed, and the goddess Athene (Minerva in Roman myth) placed in the prow a beam of wood from the oak tree at the oracle of Zeus at Dodona. Near the Bosphorous, well into their journey, the Argonauts met the aged seer Phineas, punished by Zeus for unerringly revealing the god's nefarious affairs. Whenever Phineas tried to eat, Harpies, creatures with bird's bodies and the faces of hags, swooped down to defile his food. Two Argonauts, sons of Boreas the North Wind, succeeded in driving the creatures away.

The old man told the Argonauts how to artfully navigate the clashing rocks of the Symplegades at the entrance to the Black Sea. A dove was released, and as it flew ahead the rocks rolled together, clipping its tails feathers. The way was momentaily clear as the rocks recoiled and, rowing mightily, the Argonauts just got through, losing only an ornament on the stern. After this the rocks settled, and never threatened sailors again. We can still see the dove represented in the sky by Columba (see page 140).

When the Argonauts eventually arrived in Colchis, King Aeëtes refused to hand over the Golden Fleece until Jason had completed certain tasks. Medea, the sorceress daughter of Aeëtes, fell in love with Jason, and gave him a magic potion with which to succeed in these challenges. One task was to sow a certain dragon's teeth, which then sprouted warriors, determined to kill the hero. Jason defeated them with the help of the potion. However, Aeëtes still refused to give up the Fleece, so Medea lulled to sleep the guarding dragon in the grove of Ares, and together the lovers stole away with their prize.

The complete ancient constellation of Argo Navis from a 17th-century map of the Southern-hemisphere skies by Andreas Cellarius. Argo Navis was first described by Ptolemy in the 2nd century CE.

The hero Jason about to depart on his voyage in the *Argo*, as shown in a 14th-century painting by Guido Columnis. The purpose of the trip was to retrieve the Golden Fleece from King Aeëtes of Colchis, in order that Jason could return home to reclaim his rightful throne.

CASSIOPEIA

CAS – CASSIOPEIAE / THE ETHIOPIAN QUEEN

Cassiopeia is one of the most distinctive Northern constellations, easily identified by the W-shape of her five brightest stars. She sits on the opposite side of the Pole Star to Ursa Major, culminating at midnight in early October.

Apart from immediately confirming the broad direction of the North Pole, Cassiopeia offers a precise orientation to the framework of celestial coordinates. The bright far-western star of the W, Caph (β Cas), in our era lies almost exactly on the equinoctial colure – a line running from the pole to the March equinox point (0°Aries). South of Cassiopeia, the colure passes through the Square of Pegasus, a few degrees inside the eastern edge of the Square, before reaching the equinox point (see page 109).

MAJOR STARS

α – Schedar, 2.2, yellow
The name for this star means "breast". Schedar has an unrelated companion (magnitude 8.9).

β – Caph, 2.3, white
The name derives from the Arabic title for the constellation. Caph lies 46 light years away.

γ – Cih, averaging 2.5, blue-white
This intriguing star, whose name is of uncertain origin, varies unpredictably between magnitudes 3.0 and 1.6. It is believed that its fast rotation makes it unstable, causing it to throw off rings of gas.

STARLORE

The legend of the unfortunate Queen Cassiopeia, wife of King Cepheus of Joppa, centres on the story of their daughter Andromeda (see pages 42–3). Both Cassiopeia and her daughter were beautiful. However, the queen committed a sin of hubris by declaring that they were even lovelier than the sea-nymphs, the Nereids. These were 50 charming and benevolent daughters of Nereus, the wise old man of the sea. Offended by Cassiopeia's remarks, they complained to their protector, the sea-god Poseidon (in Roman myth, Neptune). In anger Poseidon struck the waters with his trident, flooding the lands of the Palestine coast and calling up from the deep the sea-monster Cetus (in some accounts Cetus is a great whale). Cepheus consulted the oracle of Ammon to learn how to save his kingdom, and was told that his subjects could be delivered from the monster's ravages only if his daughter Andromeda was sacrificed to Cetus. The pressure of the populace was impossible to resist, and accordingly Andromeda was chained to the rocks near Joppa.

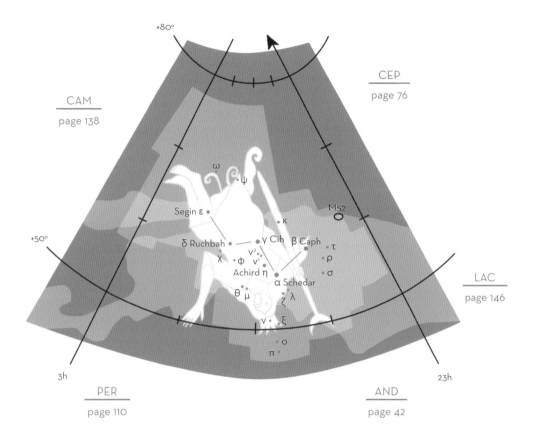

CAM
page 138

CEP
page 76

LAC
page 146

PER
page 110

AND
page 42

As Cetus began to approach the girl, Perseus chanced upon the tragic scene. He pledged to defeat the monster, but in return desired Andromeda as his wife, and Cassiopeia and Cepheus reluctantly agreed to the match. Later, at the wedding celebrations, Phineas, a jealous former suitor of Andromeda, led two hundred warriors in an attack on the happy couple, with the collusion of Cassiopeia. Perseus pulled the head of the Gorgon Medusa from its pouch and turned them all to stone.

As punishment for her vanity, Cassiopeia was cast into the heavens by Poseidon, but in a somewhat lewd and unseemly posture. As the Greak poet Aratus (3rd century BCE) comments, "No longer does she shine upon a throne ... but she headlong plunges like a diver, parted at the knees." This plunging refers to the tight circumpolar motion that raises and drops the constellation in rapid succession. Another variant recounts that she was pitched into a market-basket in which she looks absurd when tipped on her head.

THE MAJOR CONSTELLATIONS

THE ANDROMEDA GROUP

SIGNPOST MAP 1

Astriking late-autumn and winter star-field for the Northern-hemisphere observer, the Andromeda Group can be seen in this orientation at 10pm in mid-November or 8pm in mid-December.

Played out in the sky is the myth of vain Queen Cassiopeia and, northeast of her, King Cepheus, both notable circumpolar figures. To the south of them their beautiful daughter Andromeda is seen chained to the rocks, by command of the oracle of Ammon, awaiting sacrifice to the sea-monster Cetus. The monster is seen advancing toward the ecliptic from the depths of the south, separated from the rest of the group by Aries and Pisces. Brave Perseus comes in from the west on his way home with the severed head of the Gorgon Medusa. In some accounts, Perseus arrives riding the winged horse Pegasus, who shares one star on its Great Square with Andromeda. By slaying Cetus, Perseus gains the hand of Andromeda in marriage (see pages 70–1).

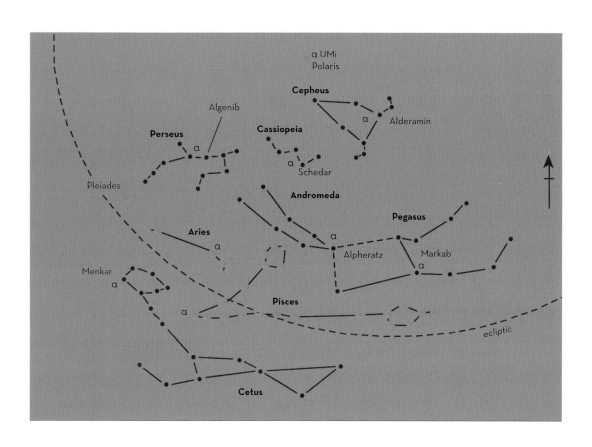

α UMi
Polaris

Cepheus

Algenib

Alderamin

α

Cassiopeia

Perseus

α

α

Schedar

Pleiades

Andromeda

Pegasus

Aries

α

α

Alpheratz

Markab

α

Menkar

α

Pisces

α

ecliptic

Cetus

CENTAURUS

CEN — CENTAURI / THE CENTAUR

Centaurus is an extensive constellation, north of Crux, on the northern shore of the Milky Way. Its centre lies approximately 50° south of Spica (α Vir, page 130), under the tail of Hydra. It is a magnificent figure of the Southern-hemisphere autumn skies, culminating at midnight during April.

MAJOR STARS

α — Rigil Kentaurus or Toliman, —0.3, pale yellow
"Centaur's foot" and "grape-vine shoot" respectively. Rigil Kentaurus is the third brightest star in the sky, and a small telescope reveals twin yellow stars of magnitudes —0.01 and 1.33, orbiting one another once every 80 years. At 4.3 light years from us, this is the closest naked-eye star to Earth after the Sun. It is believed to have been worshipped on the Nile in very early periods, with several temples oriented to its heliacal rising (its first brief appearance, as a morning star, after a period of being invisible).

β — Hadar, 0.6, blue
"Weight". The star is also known as Agena; however, the origins of both names are still uncertain.

STARLORE

For the Greek mythographers this constellation represented Chiron, the leader of the centaurs. These creatures, half-man, half-horse, had a reputation for being savage, but Chiron was an exception: he was wise and benevolent, taught humans many arts, and is credited with the design of the constellation figures themselves.

There has been confusion, however, between this Southern centaur and the zodiacal Sagittarius (see pages 116–17), who represents the aggressive type of centaur. The conflation is such that the names of the two constellations have, at times, been interchanged.

Some say that the centaurs were created when the god Cronos changed into a horse so that his wife, the Earth-goddess Rhea, would not see his coupling with Philira.

Another story concerns the Greek king Ixion, a duplicitous character invited to dine with the god Zeus (in Roman myth, Jupiter) and his wife Hera. Ixion desired Hera, who was glad to avenge her husband's adulterous dalliances. But, realizing what was happening, Zeus shaped a cloud in the image of Hera, and it was with this that the drunken Ixion mistakenly took his pleasure. Zeus surprised him in the throes of

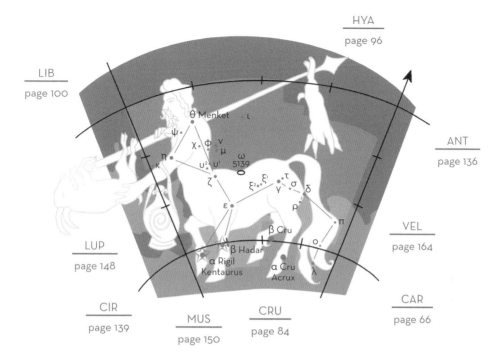

HYA
page 96

LIB
page 100

ANT
page 136

VEL
page 164

LUP
page 148

CAR
page 66

CIR
page 139

MUS
page 150

CRU
page 84

θ Menket
ι
ψ
χ φ ν
μ
η
κ
υ² υ¹
ω
5139
ζ
ξ² ξ¹
τ
σ
γ
δ
ε
ρ
π
β Cru
ο
β Hadar
α Rigil
Kentaurus
α Cru
Acrux
λ

passion and had him bound to a burning wheel, which rolls around the sky for eternity — an image of the Sun. The fake Hera took the name Nephele (meaning "cloud"); from the seed of Ixion, she bore the first centaur.

The most popular story of Chiron tells of his tragic wounding when he went to the aid of his friend Heracles (Hercules) against other centaurs. He was accidentally caught in the knee by an arrow shot by Heracles, and despite his arts and magic herbs the wound would not heal and he howled in agony. As an immortal he was fated to suffer interminably. The resolution of this miserable fate came when he was able to help Prometheus (see Aquila, pages 46–7), whose own sufferings would end only when another immortal chose voluntarily to sacrifice his status. Chiron willingly agreed to save Prometheus to grant himself the sweet release

of death. Heracles mediated and approached Zeus with the proposed transaction. Zeus assented to the exchange.

A stone relief from an archway in Sagrada San Michele, Italy, showing Centaurus holding, in this case, a hare.

CEPHEUS

CEP – CEPHEI / THE ETHIOPIAN KING

Although this constellation is not made up of bright stars, it easy to locate in the Northern hemisphere by virtue of its proximity to the pole and its distinctive shape (a square of four stars surmounted by a fifth), which makes it look like a child's drawing of a house with a steep roof. Also it stands immediately to the west of the W of Cassiopeia. It is invisible at middle-latitudes South. Cepheus culminates at midnight at the end of August.

MAJOR STARS

α – Alderamin, 2.4, white
The name comes from the Arabic for "right arm", although in modern representations the star marks the right shoulder.

β – Alfirk, 3.3, white
Meaning "flock" or "herd", this name was also sometimes given to the a star.

γ – Errai, 3.2, yellow
This star seems to herd the flock of Alfirk (β); accordingly, its name means "shepherd".

δ – averaging 3.9, yellow
The original "Cepheid variable", a supergiant varying a whole magnitude range (3.5 to 4.4) in just 5 days, 9 hours. It has an attractive blue, sixth-magnitude companion, which is resolvable with binoculars.

STARLORE

Cepheus is the father-figure in a royal family of constellations whose saga dominates the Northern skies; his queen is the vain Cassiopeia (see pages 70–1) and his daughter the beautiful Andromeda (see pages 42–3), and it is through them that he is best known. However, this constellation has a venerable history, and in Mesopotamia was identified with the king of the city-state of Babylon, who was in turn considered to be the earthly son of Bel, the Old Testament Baal and the Sumerian Enlil. This last identification gives us a clue to the astral myth of Cepheus. The Babylonians divided the heavens into three "roads", given to the cosmic trinity Ea, Enlil and Anu. Enlil held sway over the inner road, consisting of the Northern circumpolar stars, where we find Cepheus.

He has sometimes been portrayed as a regal and authoritarian figure astride the Celestial Pole, which well befits his lineage from the sky-god Enlil. As often with his treatment of the myths, the Latin poet Manilius (1st century CE) offers a detailed astrological interpretation

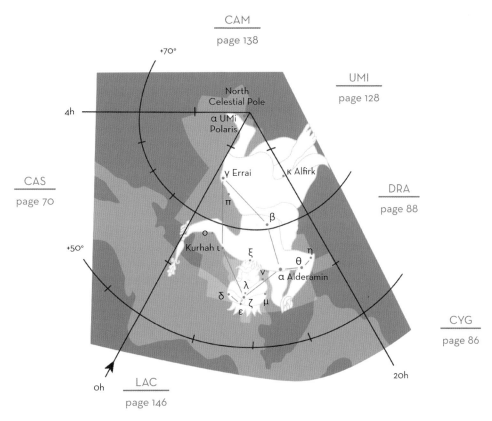

CAM
page 138

UMI
page 128

CAS
page 70

DRA
page 88

CYG
page 86

LAC
page 146

+70°

4h

North
Celestial Pole

α UMi
Polaris

γ Errai

κ Alfirk

π

β

ο
Kurhah ι

ξ

η

θ

ν

α Alderamin

λ

δ

ζ

μ

ε

+50°

0h

20h

of Cepheus in which this pole-striding king is seen as a rather self-important figure enjoying the semblance of power.

Perhaps in the end Cepheus play-acts his part without the true royal authority expected of a king. The usual treatment of him in Greek myth suggests that he is a weak character dominated by the desires of his wife. The poet Aratus (3rd century BCE) gives us the most common classical depiction of the king as "one that stretches out both his hands" — no doubt in supplication to the gods because of the ravages that the god Poseidon (in Roman myth, Neptune) had brought to his lands to punish the hubris of Queen Cassiopeia.

At one time Cepheus was identified with the Middle-Eastern storm-god Baal ("Lord"), shown here on a Babylonian stone relief.

CETUS

CET – CETI / THE SEA-MONSTER

The fourth largest constellation, Cetus sprawls along the equator with its head to the north and the main part of its body to the south. Its equatorial location ensures its visibility from most parts of the world; but its clumsy shape makes it difficult to visualize, and to find. The stars of the head, including Menkar, are south of the hind quarters of Aries; the stars of the tail, including Deneb Kaitos, are in an indistinct region below the horizontal second fish of Pisces. The midnight culmination of the stars comprising Cetus' head is at the beginning of November, but for the tail-stars midnight culmination occurs a month earlier.

MAJOR STARS

α – Menkar, 2.5, orange-red
The name for this star means "nose", although Menkar is often shown marking the monster's open jaws. This has led to its being characterized as a malevolent influence. Binoculars will reveal a sixth-magnitude, blue-white star, 93 Ceti, close by but unrelated.

β – Deneb Kaitos, 2.0, yellow
The "south part of the tail" and the constellation's lucida. It was also known to the Arabs as the "second frog"; the "first frog" was Fomalhaut (α PsA).

ζ – Baten Kaitos, 3.9, yellow
The name means "whale's belly".

o – Mira, 3.0, yellowish-red
"The wonderful one", in the neck. Roughly 23° south of Hamal (α Ari), this was the first variable star (other than novae) to be discovered, in the 17th century. It varies from third to ninth magnitude in a 332-day cycle, and on occasion flares up to second magnitude. Technically it is a "red-giant long-period variable"; similar stars are called "Mira variables".

STARLORE

Cetus is the sea-monster featured in the famous story of Andromeda (see pages 42–3), the princess who was chained to rocks as a sacrifice to the sea-god Poseidon (Neptune). While Andromeda and her parents helplessly waited for Cetus to emerge from the waters, the hero Perseus was returning from his mission to slay the Gorgon Medusa, and saw the girl. Flying with the aid of his winged sandals (or some say on the winged horse Pegasus), Perseus confused the monster by the play of his shadow on the water. Striking from above, he slashed and killed Cetus, with the magic sickle given to him by the goddess Athene.

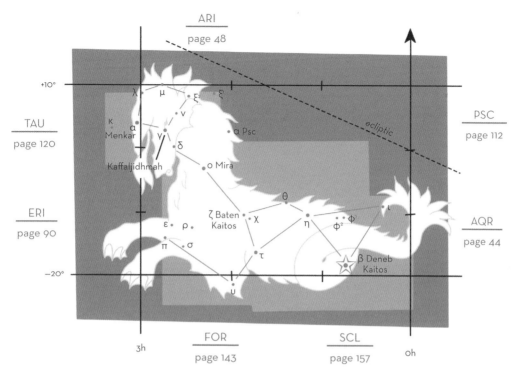

ARI
page 48

TAU
page 120

ERI
page 90

PSC
page 112

AQR
page 44

FOR
page 143

SCL
page 157

+10°
−20°
3h
0h

λ   μ   ξ²   ξ¹
κ   ν
α
Menkar
ν
δ
Kaffaljidhmah
o Mira
α Psc
θ
ζ Baten
Kaitos   χ
ε   ρ
π   σ
τ
υ
η   φ¹
φ²   ι
β Deneb
Kaitos

ecliptic

Cetus has been variously depicted as a dragon-fish or a sea-serpent, but some say that the creature is merely a great whale. As with many of the constellation figures, Cetus' classical interpretation is an overlay on archaic origins that have a wide dissemination in the cultures of the ancient Near and Middle East. In the Book of Isaiah (51:9), Jehovah hacks Rahab to pieces, and we know from Job 10.13 and 26:12 that Rahab was the sea, sometimes represented as a sea-serpent. This story in turn parallels the Babylonian creation epic, in which the sky-god Marduk flies on a white horse to slay the sea-monster and representative of primordial chaos, Tiamat. The theme of Cetus as a bringer of malice is parallelled by the indigenous peoples of northern Brazil, who saw in these stars the Jaguar, the personification of the god of thunder.

A medieval manuscript illustration showing Perseus, riding the winged horse Pegasus, about to kill Cetus and rescue Andromeda, who stands on the shore.

CRA — CORONAE AUSTRALIS / THE SOUTHERN CROWN

Corona Australis, often paired with the Northern Crown, Corona Borealis (see page 82), has no star brighter than the fourth magnitude and no named stars, but is nevertheless a distinct figure in the Tropics and the Southern hemisphere. It is completely invisible from latitudes north of 53°. On the shores of the Milky Way, it lies immediately south of Sagittarius, and its midnight culmination occurs in early July. The meteor shower of the Corona Australids occurs within this star-field every year around March 16.

STARLORE

Corona Australis is one of the original 48 constellations depicted by Ptolemy, who knew it as the "Southern wreath". Its proximity to Sagittarius explains the occasional reference to it as the Centaur's Crown, or a quiver of arrows carried by the archer.

The best-known myth relating to the figure is that of Semele, daughter of King Cadmus of Thebes. The supreme god Zeus (in Roman myth, Jupiter) disguised himself as a mortal to conduct a secret affair with the girl. To put an end to the adultery, Zeus' wife Hera transformed herself into an elderly neighbour and sowed the seeds of uncertainty in the girl's mind about her mysterious consort. Semele, by now six months pregnant, demanded that her lover reveal his identity. When Zeus refused, she did not allow him into her bed. The god then manifested himself in his full divine glory, whereupon, as Hera had predicted, the hapless maiden was consumed by a thunderbolt. The unborn infant was sewn into his father's thigh for the remaining months of his term. The child was Dionysius (Bacchus), who later braved the terrors of the underworld to retrieve his mother's soul. The gods consented to Semele's joining them on Mount Olympus, and her wreath became Corona Australis.

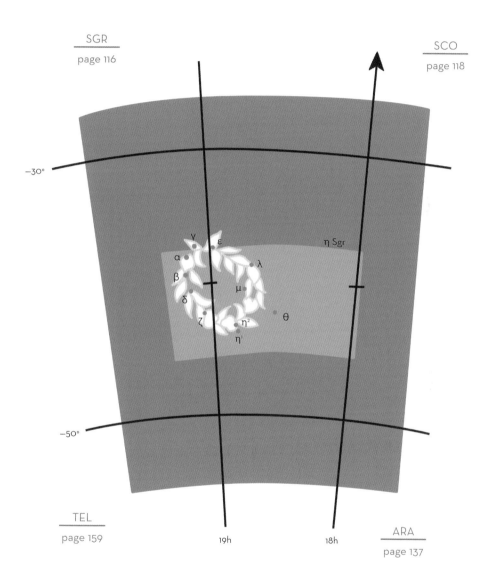

SGR
page 116

SCO
page 118

−30°

γ ε
α ε
β λ
δ μ η Sgr

ζ η²
η¹ θ

−50°

TEL
page 159

19h

18h

ARA
page 137

THE MAJOR CONSTELLATIONS

CORONA BOREALIS

CRB — CORONAE BOREALIS / THE NORTHERN CROWN

Corona Borealis is a small but distinct diadem of even stars that covers only 0.4 per cent of the sky. The ancient constellation lies between Boötes to the west and Hercules to the east. Culminating at midnight in mid-May, it is most prominent in the Northern sky in the evenings of late spring and summer.

MAJOR STARS

α — Alphecca, 2.2, blue-white
"Bright one of the dish." It was also known as Gemma, the unopened buds of a floral crown.

τ — The Blaze star, variable magnitude, pale yellow
This is an unpredictable star, which can flare up from magnitude 11 to magnitude 2.

STARLORE

In Greek myth, this constellation is the crown or wreath of Ariadne, daughter of King Minos of Crete. Every nine years, seven maidens and seven young men were sent from Athens to Crete on Minos' orders. They were offered to the Minotaur, a creature half-man, half-bull, kept in a labyrinth from which there was no escape.

Theseus, heir to the Athenian throne, presented himself as one of the seven men. In Crete, Ariadne fell in love with him and agreed to arrange his escape if he would take her back to Athens as his wife. She gave Theseus a golden ball of twine to guide him through the labyrinth. At the centre, Theseus killed the minotaur.

On the return to Athens, Ariadne was abandoned on Naxos, where she died of a broken heart, and the god Dionysius placed her wreath in the heavens. Alternatively, the constellation represents the golden thread given to Theseus.

The Persians and early Arabs knew the figure as the Dervish's Platter, the Begging Bowl or, because the circle formed by the stars is incomplete, the Broken Platter.

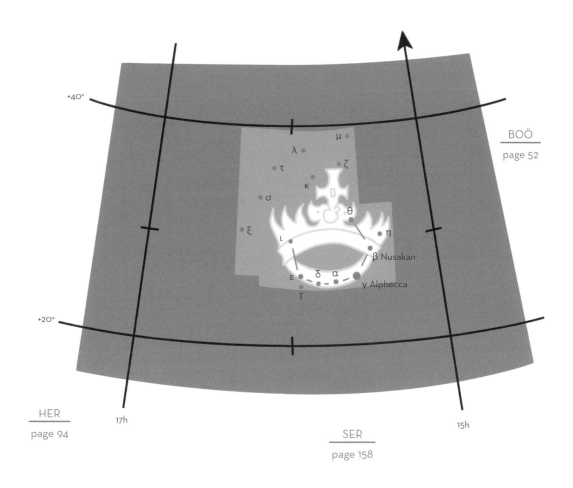

BOÖ

page 52

HER

page 94

17h

SER

page 158

15h

THE MAJOR CONSTELLATIONS

CRU — CRUCIS / THE SOUTHERN CROSS

The smallest constellation, this compact cross of four bright stars, lying across the Milky Way, makes a striking Southern figure. From middle-Southern latitudes southward to the pole, it is circumpolar (see page 16). Its midnight culmination is at the end of March. A line through its major (vertical) axis, from Gacrux (γ) to Acrux (α), points roughly toward the South Celestial Pole, some 25° away (see page 162). A line through the minor (horizontal) axis, from δ to Mimosa (β), points westward to the β and a stars of Centaurus, Hadar and Rigil Kentaurus.

MAJOR STARS

α — Acrux, 0.8, blue-white
The name, a compound of "alpha Crux", was probably coined in the early 19th century by the American astronomer Elijah Burritt.

β — Mimosa, 1.3, blue-white
This star is a Cepheid variable (see β Cep, page 76).

γ — Gacrux, 1.6, red
This star was also probably given its name by Burritt ("gamma Crux").

The Coal Sack
Lying between Acrux and Mimosa, this is a dark nebula, 400 light years away. Covering more than 30° of sky, it is easily seen silhouetted against the Milky Way.

NGC 4755 — The Jewel Box
Also known as the "κ Crucis cluster", this is a group of more than 50 stars lying 7,600 light years away. With the naked eye it appears to be a single, fourth-magnitude star.

STARLORE

The classical authors seem to identify the stars of Crux as part of Centaurus, the constellation by which it is surrounded on three sides. The Victorian scholar R.H. Allen, in his *Star Names*, drew attention to evidence of an earlier tradition that saw the cross. The 11th-century-CE Arabic astrologer al-Biruni noted that from latitude 30° North in India, a Southern asterism was visible, known as *Sula*, "the Crucifixion Beam". As Allen suggests, this may give us the clue to a reference in Dante's *Divine Comedy* (early 14th century). Passing into Purgatory at the entrance to the Southern hemisphere, Dante declares that, "setting me to spy / that alien pole, I beheld four stars / the same the first men saw, and since no living eye," (Purgatory, Canto I:22—4). The stars of Crux can no longer be seen in

CEN

page 74

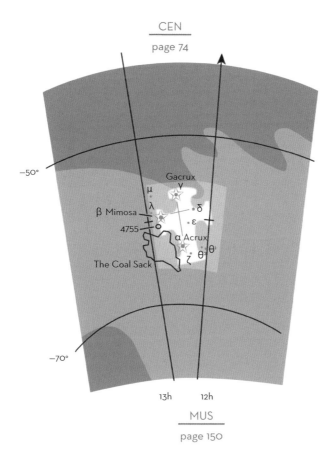

MUS

page 150

the Northern hemisphere. The "first men" are the first Christians, as Crux was just visible at the latitude of Jerusalem in the era of Christ. Dante, who was clearly aware of the effects of precession (see page 19), refers to a godless age after the death of Christ, when Crux had gradually slipped out of view at this latitude.

Crux is not seen as a cross in all cultures; in central Australia the stars have been called "The Eagle's Foot".

Dante reading from his *Divine Comedy*. His description of the cosmos, apart from being poetically beautiful, is remarkably accurate.

CYG – CYGNI / THE SWAN

Cygnus, especially notable in the evening skies of the Northern hemisphere throughout the Northern summer, has a midnight culmination at the end of July. Its figure is a swan flying away from the pole along the Milky Way. The main stars form the shape of the Northern Cross, on which the swan is located upside down. Deneb, at the bird's tail, marks the top of the cross; the swan's wings become the horizontal beam; and Albireo (β Cyg), at the bird's head, is the foot. The lucida, Deneb, with Vega (α Lyr) and Altair (αAql), form the Northern hemisphere's Summer Triangle, a striking key to orientation, visible in the evening skies from summer through to the end of the year (see page 47).

MAJOR STARS

α – Deneb, 1.3, blue-white
This star, whose name means "tail", is a supergiant, 1,700 light years from Earth.

β – Albireo, 3.0, reddish yellow
This is a beautiful double; its companion is a fifth-magnitude, blue-green star, which can be separated with binoculars. The name comes from a 16th-century mistranslation of the Arabic. The translator had thought that the allusion was to an iris flower, giving the Latin *ab ireo* meaning "from [the] iris", and this then became misspelt as Albireo.

γ – Sadr, 2.2, yellow-white
The name for this star is derived from the Arabic word meaning "breast".

NGC 7000 – The North American Nebula
This is visible with the naked eye in ideal conditions, when it appears in the shape of a bright hook in the Milky Way, 2° across at its widest. It lies approximately 1,500 light years distant.

STARLORE

The representation of the stars of Cygnus as a bird has pre-Greek origins. The original Mesopotamian figure is thought to have been known as *Urakhga*, the prototype of the Arabic *Rukh*, best known to us as the huge "roc" in the Arabian Nights' tale of Sindbad the Sailor, a fictional character based on the merchants of Baghdad.

On his second voyage, Sindbad found a roc's egg, which was 50 paces in circumference. When the parent roc arrived at the place where the egg had been found, Sindbad clung

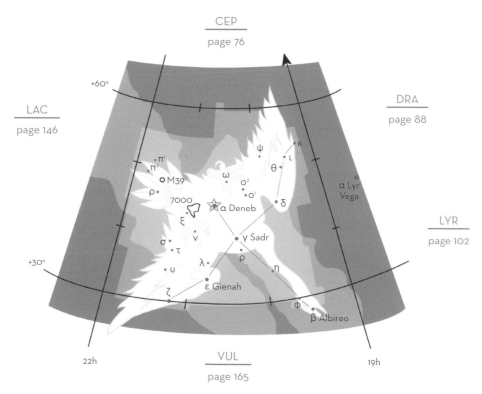

CEP
page 76

DRA
page 88

LAC
page 146

LYR
page 102

VUL
page 165

+60°

+30°

22h

19h

π¹
π²
M39
ρ
7000
ξ
σ
τ
ν
υ
ζ
λ
ε Gienah
ρ
η
☆ α Deneb
ω
O²
O¹
γ Sadr
ψ
θ
κ
ι
δ
α Lyr Vega
φ
β Albireo

to its claws and was carried to the Valley of Diamonds, so that he was able to return home with a sizeable booty of treasure.

Several Greek myths tell of young men being turned into swans. However, the best-known and most relevant tale is that of Leda, wife of King Tyndareus. Leda lay both with her husband and with Zeus (in Roman myth, Jupiter), who had taken the guise of a swan. As a result of these unions, she produced two eggs; from one she bore Helen of Troy, from the other came the Dioscuri ("sons of god"), Castor and Polydeuces (Pollux; see pages 92–3).

Leda and the Swan (c.1550), a copy of a lost painting by Leonardo da Vinci showing the mortal woman Leda with the god Zeus disguised as a swan, represented by Cygnus. The Dioscuri (the twins Castor and Polydeuces) are shown beside them.

DRACO

DRA – DRACONIS / THE DRAGON

Draco is a large, shapeless constellation, coiling indistinctly around the North Celestial Pole. Its most remarkable single feature is its head, a tight lozenge-shaped asterism of four stars, including the β and λ stars Rastaban and Eltanin. The asterism lies north and a little west of Vega (α Lyr), and north of Hercules. The midnight culmination of the head is around June 22.

MAJOR STARS

α – Thuban, 3.7, blue-white
The name comes from an Arabic term for the whole constellation. Around 2800 BCE Thuban was the Pole Star; it has now been displaced owing to precession (see page 19).

β – Rastaban, 2.4, yellow
The name means "serpent's head".

γ – Eltanin, 2.2, orange
The constellation's lucida. Its name is derived from the Arabic for "dragon's head".

STARLORE

This constellation has occasionally been seen as a snake, a hippopotamus and, in ancient India, a crocodile or alligator. Our present-day figure originated in Mesopotamia as a winged dragon, larger than the modern asterism, coiling toward the head of Ursa Major. However, the Greek philosopher Thales (fl. 6th century BCE) lopped off its wings to form Ursa Minor, and Draco has been wingless ever since.

In one story Draco represents the dragon that killed the men of Cadmus after he had sent them to the well of Ares (Mars in Roman myth) to fetch water. Cadmus slew the dragon and sowed its teeth into the earth. From the teeth sprang armed men, called "the sown men" or Sparti, the ancestors of the Thebans.

Alternatively, Draco is the dragon Ladon, slain by Heracles (in Roman myth, Hercules). Under his pledge to serve Eurystheus, Heracles was required to obtain golden apples from a tree given by the Earth-goddess Gaea to Hera on her marriage to Zeus. The tree was tended by the Hesperides, daughters of the Titan Atlas, and guarded by the eternally wakeful dragon Ladon. From the wise old man of the sea, Nereus, Heracles learned that he must not pluck the apples himself, but instead should seek help from the Titan Atlas. Heracles shot an arrow into the garden, killing Ladon and making way for Atlas to take three apples. Hera, upset about the death of Ladon, set the dragon's image in the heavens.

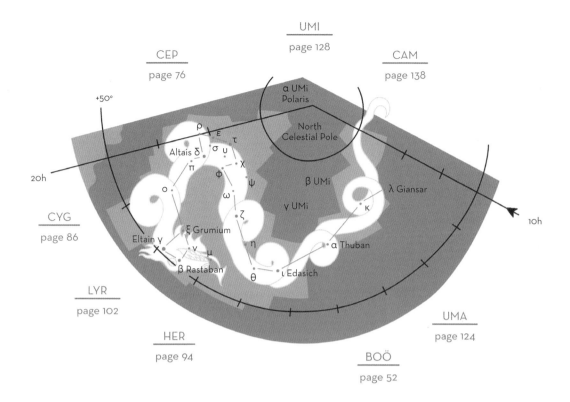

α UMi
Polaris

North
Celestial Pole

+50°

ρ
ε
τ
Altais δ
σ υ
χ
π
φ
ψ
ο
ω
β UMi
λ Giansar
20h
κ
γ UMi
ξ Grumium
ζ
Eltain γ
η
α Thuban
ν μ
β Rastaban
θ
ι Edasich
10h

Another Greek image for Draco was that of the dragon who fought for the Titans in their war against the Olympians. When 10 years of fighting had passed, the dragon faced battle with the goddess Athene (Minerva in Roman myth). She caught the beast by the tail and swung it into the sky. As it hurtled through the air, its body became knotted in itself, and was caught around the North Celestial Pole. The air was so cold here that the dragon froze in this twisted circumpolar position.

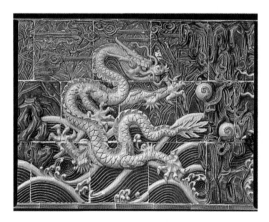

In one Chinese interpretation, at an eclipse of the Sun or Moon (represented by a pearl), the luminary is eaten by a celestial dragon. In this Chinese wall-tile illustration the dragon is about to ingest the pearl.

THE MAJOR CONSTELLATIONS

ERIDANUS

ERI – ERIDANI / THE RIVER

Eridanus is a large Southern constellation straggling from the equator to 58° South, and culminating at midnight in November. Some portion of Eridanus is always visible from any part of the world, but a full view is attained only from latitudes southward of 32° North.

Consisting of mainly faint stars, it is difficult to discern, even from the Tropics. The southern end of the river is found from first-magnitude Achernar (α Eri; 62° South), lying opposite Hadar (β Cen; 59° South) on an arc that runs through the South Celestial Pole.

MAJOR STARS

α – Achernar, 0.5, blue-white
The ninth brightest star in the sky. Its name comes from the Arabic for "end of the river", although this designation was originally applied to θ Eri.

β – Cursa, 2.9, blue-white
The name means "throne", or "foot-stool", and refers to the star's position close to Orion.

γ – Zaurak, 3.0, yellow-red
The name means "boat".

STARLORE

The stars of this constellation have been associated with various Earthly rivers, including the Euphrates, the Nile and the Padus (the Po in Italy). Aratus (3rd century BCE) was the first classical author to name the constellation Eridanus, although he may have drawn upon an earlier Mesopotamian designation.

Aratus refers to "those poor remains of Eridanus, river of many tears". He is alluding to the idea that the river was partly burnt up, no doubt to explain the feebleness of its stars, as a consequence of the tragic story of Phaethon. This youth, whose name means "shining", was the mortal offspring of the Sun-god Helios and the Oceanid (sea-nymph) Clymene. Desiring to

establish the truth of his parentage, Phaethon came to the Sun-god's palace. Here, Helios acknowledged that he was indeed Phaethon's father, and to prove it he promised to satisfy any wish requested by his son that was within his power. It was a fateful oath because, despite his father's protestations, Phaethon demanded to be allowed to drive the Chariot of the Sun for a single day. As Helios and his son began their swift ascent, Phaethon lost control of the chariot and the horses bolted from the road of the Sun, bumping into the constellations and plunging to the depths of the sky. As the chariot passed close to the Earth, it set the mountain peaks alight, and fires raged down the slopes

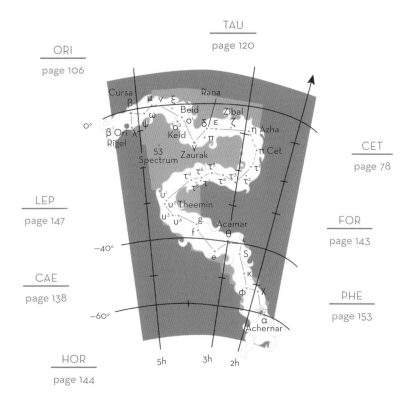

ORI
page 106

TAU
page 120

CET
page 78

LEP
page 147

FOR
page 143

CAE
page 138

PHE
page 153

HOR
page 144

into the valleys below, charring the earth and drying up all the rivers. The Earth-goddess Gaea let out a shriek of horror and Zeus (in Roman myth, Jupiter) intervened to save the world from destruction. He hurled a thunderbolt at the chariot, driving the maddened horses down toward the sea. The blazing body of Phaethon fell into Eridanus, and the water quenched the flames. The Naiads (freshwater nymphs) and the daughters of Helios came to mourn; they shed tears which turned to amber, and themselves became poplars on the river bank.

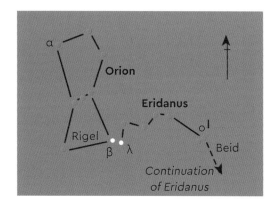

The Northern part of the river starts close by Orion's feet. A few degrees northwest of β Ori (Rigel) lies Cursa (β Eri). From here trace out the path of the river southward.

GEMINI

GEM – GEMINORUM / THE TWINS

Gemini, the third zodiacal constellation, lies northeast of Orion. The brightest stars, Castor and Pollux, lie 4¹/₂° apart, some 30° north of Procyon (α CMi), and mark the heads of the twins. Gemini is a fine sight in the Northern winter skies, culminating at midnight in January.

At middle-latitudes South during Southern summer, the twins appear low in the sky. The constellation resembles a rectangle lying slantwise across the ecliptic. Its third bright star Alhena (γ Gem) marks the feet of the twins, who paddle in the Milky Way.

MAJOR STARS

α – Castor, 1.6, blue-white

This is a remarkable compound system of six stars, comprising three pairs of binaries (any system beyond six stars is thought to become unstable). The group lies 47 light years distant. The dual composition of Castor symbolizes the perpetual duality associated with Gemini.

β – Pollux, 1.1, yellow

This star is the constellation's lucida. Ovid (43 BCE–17 CE) termed it Pugil, the pugilist of the two brothers. Early Arabian astronomers called it Rasalgeuse, meaning "head of the twin".

γ – Alhena, 1.9, blue-white

The name for this star may derive from the Arabic for a brand on a horse or camel (the stars γ, μ, ν, η and ξ were seen as a camel's hump).

M35

This star cluster of 200 stars, 2,800 light years distant, is visible to the naked eye as a fifth-magnitude patch, roughly the same size as the Full Moon.

STARLORE

The association of the two brightest stars of Gemini with an earthly pair has been universal. In Egypt they were a pair of sprouting plants, and they have been identified in Phoenician culture as a pair of kid goats. The Mesopotamian prototype shows them as naked twin boys. In one Roman interpretation they were Romulus and Remus, the legendary founders of Rome.

In Greek myth the twins are Castor and Polydeuces (Pollux to the Romans), the Dioscuri ("sons of god"). They were born from an egg laid by Leda, queen of Sparta, after she had coupled with Zeus (Jupiter) disguised as a swan (see pages 86–7). Mortal Castor was the son of Leda's husband; immortal Polydeuces was the son of Zeus.

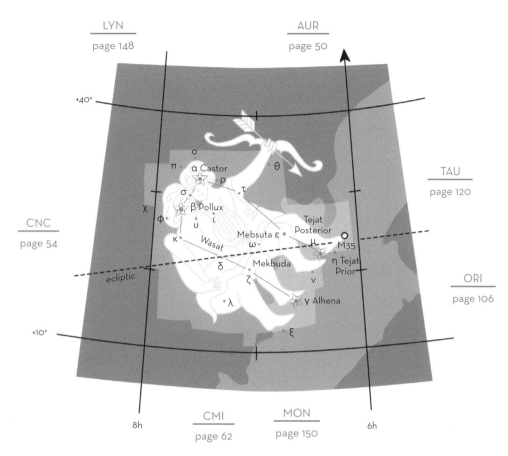

LYN
page 148

AUR
page 50

TAU
page 120

CNC
page 54

ORI
page 106

CMI
page 62

MON
page 150

+40°

o
π · α Castor
ρ
θ
σ · τ
χ
φ · β Pollux
υ · ι
Tejat
Posterior
κ · Wasat
Mebsuta ε
ω · μ · M35
δ · Mekbuda
η Tejat
Prior
ζ
ν
λ
γ Alhena
ξ

ecliptic

+10°

8h
6h

The twins travelled to the land ruled by Idas and Lynceus. Idas killed Castor with a spear, whereupon Polydeuces, although injured, killed Lynceus. Zeus intervened and struck Idas dead. Polydeuces refused to accept his immortality unless Castor could share it. Zeus allowed the two to alternate their days for ever between the realm of the gods and the underworld, Hades.

Poseidon (Neptune) made the twins the protectors of sailors; they were both members of the crew (the Argonauts) that Jason enlisted to help him retrieve the Golden Fleece. Accordingly, the stars Castor and Pollux stand high above the mast of Argo Navis (see pages 66–7).

Gemini, from a medieval manuscript. The twins are often represented with their arms entwined, owing to the proximity of their α stars.

THE MAJOR CONSTELLATIONS

HERCULES

HER – HERCULIS / THE HERO

Despite the mythical importance of its hero, the Northern constellation Hercules is not, at first sight, very impressive. It is a large but scattered figure, with no stars brighter than third magnitude. It can be located from Vega (α Lyr), which lies immediately to its east. Hercules culminates at midnight in June.

MAJOR STARS

α – Ras Algethi, averaging 3.5, red
The name derives from the Arabic for "head of the kneeler". One of the largest stars known, Ras Algethi is a red giant, 600 times the diameter of the Sun.

β – Kornephoros, 2.8, pale yellow
This star is the lucida of the constellation.

STARLORE

Hercules presented a mystery for the classical authors as a strange "kneeling man". The association of the figure with Heracles, the great hero of Greece, dates from the 5th century BCE. Behind him stands Gilgamesh, the central character of the Babylonian *Epic of Creation*. From the late 4th millennium BCE, this ancient progenitor appears resting on one knee, his foot on the head of a dragon. This is an exact image of the constellation Hercules, with his foot on the head of Draco.

In Greek myth, Heracles is immensely brave and good-hearted, but rash. His mother was a mortal, who unknowingly coupled with Zeus (Jupiter, in Roman myth). Hera, Zeus' jealous wife, vowed to kill Heracles. She wreaked her most terrible revenge when she drove him temporarily insane. During this fit he killed his wife Megara and their three sons. The noble Theseus befriended him after the tragedy and took him to Athens, but he could not be rescued from his profound sense of guilt. He sought help from the oracle at Delphi, and was advised that he could purge himself only if he carried out a penance. He should go to his cousin Eurystheus, king of Mycenae, and submit to whatever his cousin demanded. Eurystheus set Heracles 12 near-impossible tasks, the so-called Labours of Heracles. These were (in order of accomplishment) to slay the Nemean lion (see page 99) and the Lernean Hydra (see page 98); to capture the Cerynean hind and the Erymanthian boar; to clean the Augean stables; to kill the Stymphalian birds; to capture the Cretan bull and the flesh-eating mares of Diomedes; to seize the girdle of Hippolyte,

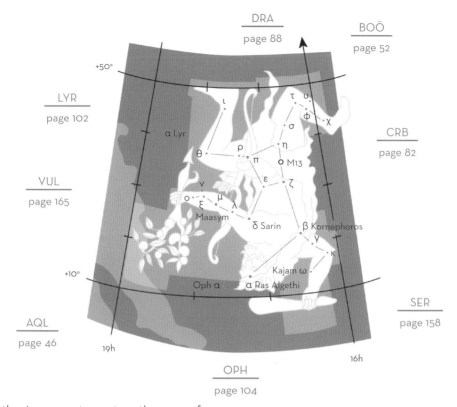

DRA
page 88

BOÖ
page 52

LYR
page 102

CRB
page 82

VUL
page 165

SER
page 158

AQL
page 46

OPH
page 104

+50°

+10°

19h

16h

α Lyr

ι

τ υ
φ χ
σ
η

θ ρ
π   ο M13
ε ζ

ν

ο ξ μ λ
Maasym

δ Sarin   β Kornephoros
γ
κ

Kajam ω

Oph α   α Ras Algethi

queen of the Amazons; to capture the oxen of the three-bodied Geryones; to steal the golden apples of the Hesperides; and, finally, to raise the hell-hound Cerberus from the underworld.

Heracles died after his wife Deianeira was tricked into thinking that a deadly poison (given to her by the centaur Nessus) was a harmless potion to guarantee her husband's fidelity. She covered Heracles' shirt with the poison and it ravaged the hero's body. As he lay on his funeral pyre, Heracles was borne up to Mount Olympus on a cloud. There he was reconciled with Hera and granted immortality by Zeus.

A cylinder seal dating from 1400–1300 BCE showing the Babylonian hero Gilgamesh with his companion Enkidu. It is thought that Gilgamesh was the precursor of the Greek hero Heracles (represented by Hercules in the sky).

HYDRA

HYA — HYDRAE / THE WATERSNAKE

This is the largest constellation, stretching for more than 100° of sky. Portions of Hydra are visible from anywhere in the world, but the predominence of fourth- and fifth-magnitude stars make its long body difficult to discern. Its lucida is Alphard (α Hya), but its most distinctive feature is the delicate group of six stars at the head of the snake, 15° east of Procyon (αCMi). For the observer at or above middle-latitudes North, Hydra appears low above the horizon and therefore requires ideal viewing conditions (see pages 20–1). The six stars in the head culminate at midnight around January 31; the tail portion, some 13° south of Spica (α Vir), comes to midnight culmination in April.

MAJOR STARS

α – Alphard, 2.0, orange
The name for this star means "solitary", and is derived from the Arabic for "solitary one in the Serpent", which suits its position as the only bright star in this area of sky.

STARLORE

The two small constellations riding on the snake's back, Corvus and Crater, invite us to distinguish different portions of the water snake, and mythographers frequently divided Hydra accordingly. The astronomer John Flamsteed (1646–1719) saw four divisions: from head to tail or west to east, these are Hydra, Hydra and Crater, Hydra and Corvus, and the Continuation of Hydra. The figure as a whole is considered to be a female watersnake, in contrast with the modern Southern watersnake Hydrus, which is male, and was created to complement Hydra.

Hydra is an ancient constellation. There is some Mesopotamian evidence, dating from c.1200 BCE, that it was identified with the primeval watersnake Tiamat, slain by Marduk in the great war of the gods. However, this association has also been found for Draco, the dragon (see pages 88–9), and Cetus, the sea-monster (see pages 78–9).

The best-known tale relating to this snake identifies it with the Lernean Hydra challenged by Heracles (Hercules) in the second of his 12 Labours. Lerna, a fertile and sacred coastal region near the city of Argos, became terrorized by the monstrous Hydra, which inhabited a swamp of unknown depth. The creature had the body of a dog and (most commonly) nine heads, each breathing venomous fumes. It could sprout two or three new heads whenever one was lopped off or crushed. To defeat the monster, Heracles followed the strategy

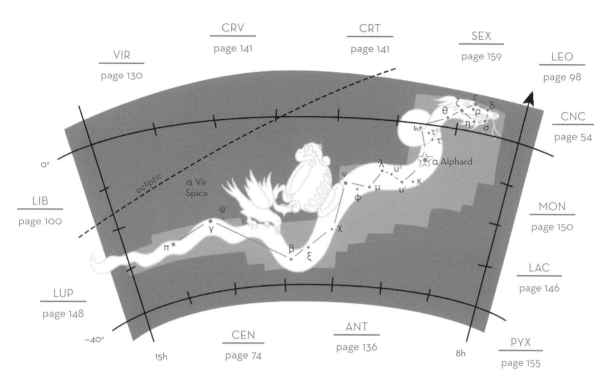

VIR
page 130

CRV
page 141

CRT
page 141

SEX
page 159

LEO
page 98

CNC
page 54

LIB
page 100

MON
page 150

LAC
page 146

LUP
page 148

CEN
page 74

ANT
page 136

PYX
page 155

advised by Athene: he forced it out of its lair by shooting burning arrows at it, and held his breath while he closed in for combat. He chopped off the heads, but more grew with every blow. At that moment a crab sent by Hera scuttled out of the swamp and nipped Heracles' foot, but it was promptly crushed and became the constellation Cancer (see page 54).

Heracles' charioteer Iolaus now came to his aid. Iolaus set fire to a corner of the grove and, taking up blazing branches, scorched and cauterized the wounds where the heads of the creature had been severed, thus staunching the flow of blood and preventing new heads from forming. At the same time Heracles found the golden, immortal head of Hydra among the seething mass; he struck it off and buried it under a heavy rock. He disembowelled the body and dipped his arrows into the creature's bile. From then on, any wound from one of those arrows was fatal.

Hydra, with Crater (left) and Corvus (right) riding on the watersnake's back. From an 11th-century manuscript edition of Ptolemy's star lists, originally published in the 2nd century CE.

LEO — LEONIS / THE LION

Leo is the fifth zodiacal constellation, and certainly the most easily recognizable figure: a crouching lion facing westward, its distinctive head and mane marked by a sickle of stars like a reversed question mark, curving northward from Regulus (α Leo). Leo lies south of the "pointer" stars in the Plough or Big Dipper (see Ursa Major, pages 124–5) and to the northwest of Virgo. Its midnight culmination is around March 1.

MAJOR STARS

α – Regulus or Cor Leonis, 1.4, blue-white
The names for this star mean, respectively, "little king" and "lion's heart". The star lies on the ecliptic, and was the leader of the four Royal Stars or heavenly "watchers" of Mesopotamia, the other three being Aldebaran (α Tau), Antares (α Sco) and Fomalhaut (α PsA).

β – Denebola, 2.1, white
The name Denebola means "lion's tail".

γ – Algieba, 1.9, orange-yellow
This star's name means "forehead". Algieba is a binary system consisting of two giant stars.

STARLORE

The lion has been identified with the Sun since the early civilization of Mesopotamia. The Egyptians connected Leo with the heliacal rising of Sirius (see pages 58–61), and with the rise of the Nile in high summer, as these events coincided with the Sun's passage through Leo. The constellation's link with the Nile is a possible explanation of why Greek and Roman architects often placed a lion's head at springs and fountains; this architectural device can be traced to the Egyptians who placed a lion's head at their canal gates.

In the 12 Labours of Heracles (in Roman myth, Hercules), Leo is identified with the Nemean lion. Heracles was required to skin a monstrous lion whose pelt was impervious to stone or metal. Having wrestled it with his bare hands and choked it to death, he used the beast's own claws to skin it. He took the pelt as a cloak of invulnerable armour, and donned the lion's head as a helmet.

Leo is also said to be the lion in the tragic story of the lovers Pyramus and Thisbe. In his *Metamorphoses*, the Roman poet Ovid (43 BCE–17 CE) tells how their parents forbade their union. They talked secretly through a chink in the wall between their houses, and one day made a plan to meet outside the city beside a certain mulberry tree with white berries. When Thisbe came to the meeting-place,

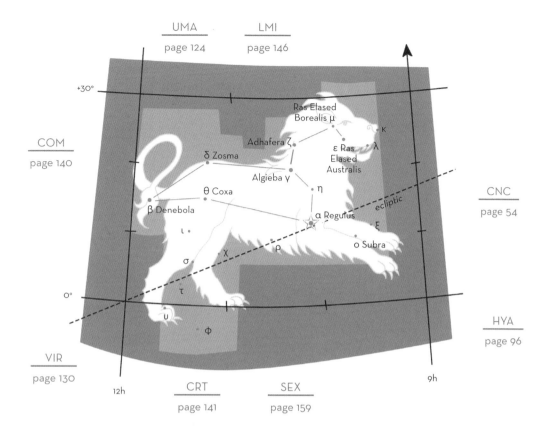

UMA
page 124

LMI
page 146

COM
page 140

CNC
page 54

HYA
page 96

VIR
page 130

CRT
page 141

SEX
page 159

+30°

Ras Elased
Borealis μ

Adhafera ζ

ε Ras
Elased
Australis

κ

λ

δ Zosma

Algieba γ

η

θ Coxa

β Denebola

α Regulus

ecliptic

ξ

ι

ο Subra

χ

ρ

σ

τ

υ

φ

0°

12h

9h

Pyramus was not there, but she was startled by a lion, bloody from a kill. As she ran away her veil slipped and fluttered past the lion, which snatched it with its paw. When Pyramus arrived he saw Thisbe's torn veil, bloody from the lion's paw, and assumed that his love had been eaten. In anguish he killed himself with his sword. At this moment, Thisbe ran back and flung herself on her dead lover's body, before taking the sword and thrusting it into her own flesh. The lovers' blood coloured the white mulberries red — their colour ever since. To remind parents not to deny young love, Zeus placed the veil in the heavens as Coma Berenices (see page 140), floating down by the lion.

An ancient gold coin showing a lion in front of the Sun. The lion has taken on solar imagery since the civilization of ancient Mesopotamia.

THE MAJOR CONSTELLATIONS

LIBRA

LIB – LIBRAE / THE SCALES

Libra, the seventh zodiacal constellation, lies between Virgo to the west and Scorpius to the east. Visually unassuming, it can be most easily identified from Scorpius, by extending the scorpion's small pincers to form a huge pair of claws. The scales' fulcrum, Zuben Elgenubi (β Lib), lies almost exactly on the ecliptic, midway along and a few of degrees north of a line between Spica (α Vir) and Antares (α Sco). Libra culminates at midnight in early May; it is visible at all latitudes outside the Arctic region.

MAJOR STARS

α – Zuben Elgenubi, 2.8, blue-white
A double star, resolvable through binoculars. α1 (magnitude 5.2) is a white companion star to α2 (2.8). The name comes from the Arabic for "southern claw", a reminder of the Greek tradition that Libra was an extension of the claws of Scorpius.

β – Zuben Eschamali, 2.6, emerald green
The name means "northern claw". Its distinct green tinge is rare among stars.

STARLORE

The Greeks usually mingled the stars of Libra into Scorpius, but the image of the scales was recognized, and there may be a Mesopotamian origin for this symbolism. Latin authors treated Libra as fully distinct from Scorpius. The balance also symbolized the equal lengths of day and night at the equinoxes: two millennia ago, the Sun passing into Libra marked the September equinox. From Roman astrologers came the idea that the scales are those of Justice, held by Astraeia, goddess of justice (neighbouring Virgo).

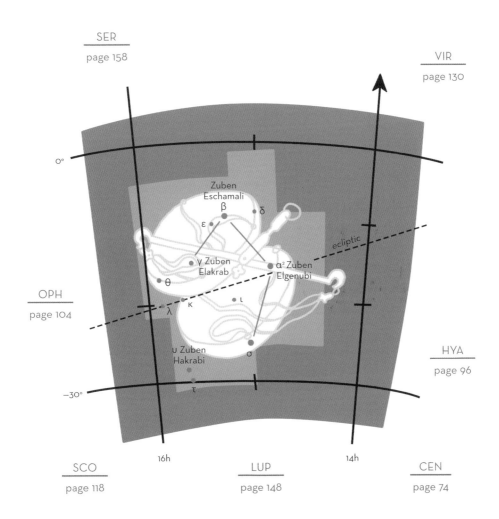

SER

page 158

VIR

page 130

0°

Zuben
Eschamali
β

δ

ε

ecliptic

γ Zuben
Elakrab

α² Zuben
Elgenubi

θ

OPH

page 104

λ

κ

ι

υ Zuben
Hakrabi

σ

HYA

page 96

−30°

τ

16h

14h

SCO

page 118

LUP

page 148

CEN

page 74

THE MAJOR CONSTELLATIONS

LYRA

LYR — LYRAE / THE LYRE

This beautiful constellation lies on the western edge of the Milky Way. It is easy to spot owing to its brilliant lucida Vega (α Lyr), which forms one corner of the Summer Triangle. The other stars in the triangle are Deneb (α Cyg) and Altair (α Aql). Lyra, impressive from the Northern hemisphere and Tropics, begins to disappear from view at middle-Southern latitudes: Vega can just be made out on the horizon at 50° South. Lyra culminates at midnight in early July.

MAJOR STARS

α — Vega, 0.03, blue-white
The fifth brightest star in the heavens, at a distance of 26 light years. In the North it is second only to Arcturus (α Boö). Its name derives from the Arabic for "swooping vulture (or eagle)"; to the Arabs Lyra was a bird with half-closed wings, thought to derive from ancient Indian starlore.

M57 — The Ring Nebula
Lying at 2,000 light years distance, M57 can be seen through a small telescope as an elliptical hazy disk.

STARLORE

For the Greeks Lyra was the instrument invented by the infant Hermes and given by Apollo to his son, Orpheus, who went into the underworld to seek his bride Eurydice: she had been killed by a viper. Hades, king of the underworld (Pluto in Roman myth), was touched by his music and permitted Orpheus to take Eurydice back, provided that he did not turn to look at her until they had emerged from hell. However, at the very last moment Orpheus glanced back, and Eurydice's soul slipped away for ever.

DRA
page 88

CYG
page 86

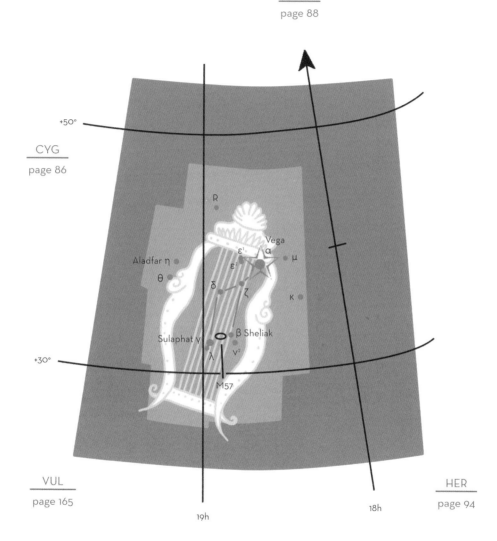

+50°

R

Vega
ε¹
α
Aladfar η   ε²   μ
θ

δ   ζ
κ

Sulaphat γ   β Sheliak
λ   ν²
+30°

M57

VUL
page 165

HER
page 94

19h   18h

THE MAJOR CONSTELLATIONS

OPHIUCHUS

OPH – OPHIUCHI / THE SERPENT HOLDER

Ophiuchus straddles the equator south of Hercules. The serpent that he carries falls at either side of him – to the west its head (Serpens Caput) and to the east its tail (Serpens Cauda). Ophiuchus' eastern leg plunges through the Milky Way onto the ecliptic and due south toward the distinctive figure of Scorpius. In this frozen moment Ophiuchus is seen grinding the scorpion into the dust. Cuminating at midnight in early June, the constellation is fully visible in the Northern summer and the Southern winter skies for latitudes between 60° North and 76° South. However, this is a faint figure, requiring patience to draw out the shape. The most recent supernova explosion in our Galaxy was observed in Ophiuchus in 1604, and recorded by Johannes Kepler; at maximum brightness its apparent magnitude was around –3.

MAJOR STARS

α – Ras Alhague, 2.1, white
The name is derived from the Arabic for "head of the serpent-charmer". Ras Algethi (α Her) at the head of Hercules lies just 6° west and a little north.

β – Cebalrai, 2.8, yellow
The name is believed to derive from the Arabic for "heart of the dog". The early Arabs saw in this area a shepherd and his flock, and a dog at Ras Algethi (α Her).

δ and ε – Yed Prior and Yed Posterior, 2.7 and 3.3, both orange
Both these stars are giants. They mark Ophichus' hand. As a useful benchmark for observation, δ is exactly half a magnitude brighter than ε.

STARLORE

The serpent-holder and the serpent that coils around him were in early times seen as a single constellation. "The struggle," says the poet Manilius (1st century CE), "will last for ever as they wage it on level terms with equal powers." In Greek, Ophiuchus means "toiling"; but there is no classical hero of this name, and the figure was thought to represent the legendary healer Aesculapius, believed to be the ancestor of Hippocrates (born c.460 BCE), the great physician of Cos. The universal symbol of medicine belongs to him: the caduceus (a staff with two intertwined serpents).

Aesculapius' story is as follows. His mother Coronis was wooed by the god Apollo, who left a white crow to keep an eye on her. However, Coronis desired a man called Ischys, and although she was already pregnant by Apollo, she lay with

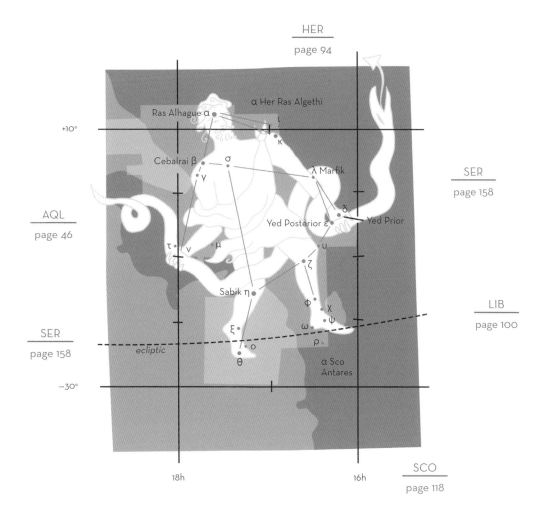

HER
page 94

α Her Ras Algethi

Ras Alhague α ι

κ

SER
page 158

Cebalrai β σ

γ λ Marfik

δ

AQL
page 46

Yed Posterior ε Yed Prior

+10°

τ ν μ υ

ζ

Sabik η

φ χ
ψ

ω

LIB
page 100

ρ

SER
page 158

ξ o
ecliptic

θ

α Sco
Antares

−30°

18h 16h

SCO
page 118

her mortal lover. The crow set out with its report, but Apollo had already divined the act and was furious that the crow had not pecked out Ischys' eyes. He turned the creature black with a curse, and crows have been black ever since.

Apollo complained of the infidelity to his sister, the huntress Artemis, who unleashed a quiver of arrows at Coronis. It was only when the corpse of Coronis lay on the funeral pyre that Apollo felt remorse. There was nothing to be done for Coronis, but Hermes stepped in to cut the unborn child from its mother's womb; thus Aesculapius was saved. He was given into the care of the kindly centaur Chiron (see Centaurus, pages 74–5), who taught him all the arts of medicine. However, his ability to raise the dead threatened the realm of the underworld, so its king, Hades (in Roman myth, Pluto), protested to Zeus (Jupiter), who struck Aesculapius dead with a thunderbolt. In revenge, Apollo killed the Cyclopes, who made Zeus' thunderbolts.

THE MAJOR CONSTELLATIONS

105

ORION

ORI – ORIONIS / THE HUNTER

With its brilliant array of first- and second-magnitude stars and its distinctive pattern, Orion heads the first rank of constellations. Only the Plough or Big Dipper in the North, or Crux in the South, could compete for instant recognition. To see the mighty giant with spangled belt for the first time after several months' absence, striding high, marks a turning point each year, as he ushers in the Northern autumn or the Southern spring. Orion culminates at midnight in mid-December when it is fully visible from all places outside the Arctic and Antarctic regions. Finding Orion from other stars is to reverse the order of things; even so we can describe the figure as lying partly on the Milky Way, south of Auriga (see pages 50–1) and west of Procyon (α CMi; see page 62). With Procyon and Sirius (α CMa; see page 58), Orion's α star, Betelgeuse, forms a notable triangle, with the apex at Sirius facing due south. The line of three bright stars in Orion's belt points directly toward Sirius.

MAJOR STARS

α – Betelgeuse, 0.5, red

This slightly variable star is a supergiant lying at a distance of 425 light years. It has a diameter 300–400 times that of our Sun. Its name, rather prosaically, derives from the Arabic for "armpit of the central one".

β – Rigel, 0.12, blue-white

"Giant's leg". The seventh brightest star in the sky and Orion's lucida, it is a supergiant more than 1,000 light years distant, and makes a distinctive colour contrast with Betelgeuse.

γ – Bellatrix, 1.6, pale yellow

The name means "female warrior", which derives from a rather loose medieval translation of the Arabic Al Najid, meaning "the conqueror".

M42

This deep-sky object can be located around θ1 and θ2 Ori. It is the finest diffuse nebula in the sky, clearly observable to the naked eye as a hazy patch of 1° square. The nebula is 1,500 light years distant and around 15 light years in diameter. Within it a star cluster is in the process of formation.

STARLORE

In Greek myth Orion was a hunter of great prowess. To show off his skill, he foolishly boasted that he could kill all living beasts. The Earth-goddess Gaea, alarmed at Orion's claim, sent a scorpion to kill him. The myth is borne out in the night sky. As the stars of Scorpius

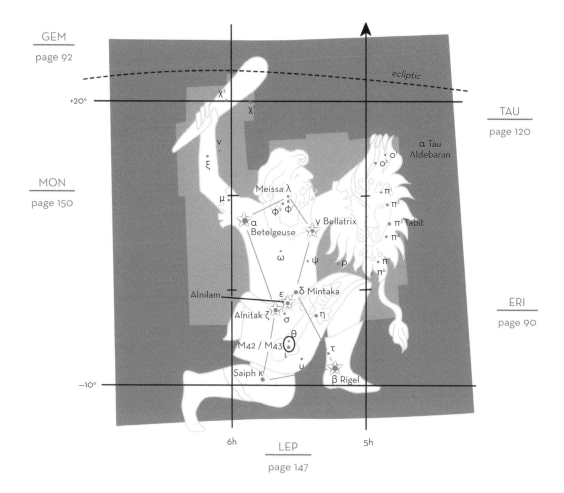

ecliptic

+20°

χ²

χ¹

ν

ξ

α Tau
Aldebaran

o¹

o²

μ

Meissa λ

φ² φ¹

α
Betelgeuse

γ Bellatrix

π¹

π²

π³ Tabit

π⁴

ω

ψ

ρ

π⁵

π⁶

Alnilam

ε

δ Mintaka

Alnitak ζ

σ

η

θ

M42 / M43

ι

τ

υ

Saiph κ

β Rigel

−10°

6h

5h

(see pages 118–19) rise in the east, Orion sinks defeated in the west. The story is completed when, as Scorpius sets in the west, the creature is crushed into the earth by the healer Aesculapius (Ophiuchus; see pages 104–5), who, having killed it, revives Orion so that he soon rises once more in the east, completely renewed.

This theme of death and rebirth could be a legacy of a far more ancient conception. Five of the Fourth-dynasty pyramids on the Giza Plateau in Egypt are positioned in an earthly representation of Orion, with the River Nile mirroring the Milky Way. The southern sighting shaft of the Great Pyramid is aligned to the belt stars of Orion, and in particular ζ Ori, as they would have been in 2700–2600 BCE. In this period our figure Orion represented the Egyptian god of the dead and the first king of Egypt, Osiris. After death, the pharaoh was thought to mystically inseminate these stars to ensure that, through Osiris, Horus the Sun-god would be reborn in his successor.

PEG — PEGASI / THE WINGED HORSE

Pegasus is visible as far down as middle-latitudes South, culminating at midnight in September. The main body comprises the distinctive Great Square, including the α-star of Andromeda. The east side of the Square approximately marks the equinoctial colure (a line through the poles and the two equinox points); the west side gives a north–south line that crosses the equator and passes near Fomalhaut (α PsA; see page 114).

MAJOR STARS

α Andromedae — Alpheratz, 2.1, blue-white
The northeastern corner of the Great Square. Once δ Peg, it was also known under the name *Sirrah* ("navel"), marking the navel of the horse. See also Andromeda, pages 42–3.

α — Markab, 2.5, blue-white
The name means "saddle", implying anything that bears a rider. The star marks the southwestern corner of the Great Square.

β — Scheat, averaging 2.5, deep yellow
This name means "shin", but the star is also sometimes called *Menkib* ("shoulder"). Scheat sits at the northwestern corner of the Great Square.

γ — Algenib, 2.8, blue-white
This star, whose name means "side", lies at the southeastern corner of the Great Square. It is a pointer to the March equinox point, roughly 15° south of Algenib (see diagram, opposite).

ε — Enif, 2.4, yellow
A supergiant, whose name means "nose". Good binoculars should reveal a blue companion-star of the eighth magnitude.

ζ — Homam, 3.6, white
The name may be derived from *Sa'd al Humam* ("lucky star of the hero"), referring to the man who could mount the divine steed. The star has also been called *Al Hammam* ("whisperer"), which could refer to the secret art of "horse-whispering", practised in ancient times among Romanies, whereby a wild horse may be tamed by gentle contact with a trainer who imitates the creature's movements.

η — Matar, 2.9, yellow
From *Al Sa'd al Matar*, meaning "fortunate rain".

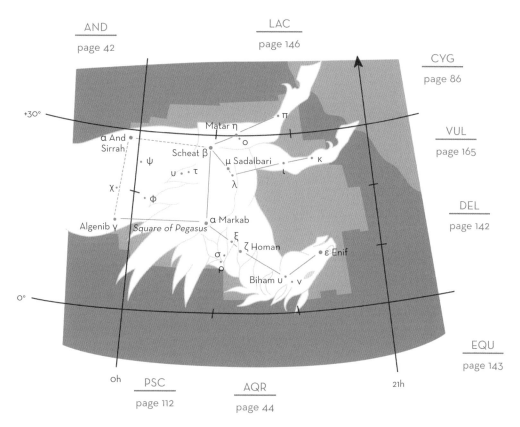

AND
page 42

LAC
page 146

CYG
page 86

VUL
page 165

DEL
page 142

EQU
page 143

PSC
page 112

AQR
page 44

+30°

Matar η

α And
Sirrah

ο

π

Scheat β

μ Sadalbari

κ

ψ

ι

υ τ

λ

χ

φ

Algenib γ   *Square of Pegasus*

α Markab

ξ

ζ Homan

ε Enif

σ

ρ

Biham υ

ν

0°

0h

21h

STARLORE

Pegasus is usually depicted as a winged horse, a tradition originated in Mesopotamian and Etruscan starlore. He was conceived when the sea-god Poseidon (Neptune to the Romans) disguised himself as a horse to seduce the Gorgon Medusa. When Perseus killed Medusa, Pegasus sprang up from her body, fully formed.

Pegasus was the steed of the hero Bellerophon. In a dream, Athene came with a golden bridle and advised Bellerophon to ride Pegasus. Subsequently, Pegasus allowed Bellerophon to mount him. Some say that Perseus rode the horse in his rescue of Andromeda (see page 110).

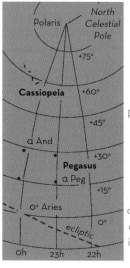

The left-hand (eastern) edge of the Square of Pegasus roughly indicates 0h of Right Ascension (the equinoctial colure). The top and bottom edges of the square show +30° and +15° of declination. Duplicate this quare immediately below itself to find 0° Aries, the March equinox point.

PERSEUS

PER – PERSEI / THE HERO

A Northern constellation to be found partly on the Milky Way, Perseus lies between Cassiopeia to the northwest and Taurus beneath his feet to the south. It culminates at midnight in November and is a fine sight in the Northern winter skies, but is lost from view at middle latitudes in the Southern hemisphere.

In this constellation the Perseids, the most striking meteor shower of the year, peak on August 12 or 13 when they flash out from their radiant or focal point near λ Per at a rate of 60–70 meteors every hour (see also page 24). Evidence of the shower can be seen for several weeks before and after the peak date.

MAJOR STARS

α – Mirfak, 1.8, brillliant yellow

The name for this star means "elbow". It is also known as Algenib ("side"), but this can lead to confusion with the star of the same name in Pegasus (γ Peg). A loose cluster of stars, known as Melotte 20, can be seen with good binoculars in the vicinity of Mirfak.

β – Algol, 2.1, white

The demon-star, literally "the ghoul". It has been noted as red in colour. This is among the most notable stars for both astronomers and mythographers. It is a defining example of the "eclipsing binary", the first to be discovered. The phenomenon occurs where two nearby stars form a single system around a shared centre of gravity, periodically eclipsing each other and thus reducing the apparent magnitude of their light. Algol "blinks" every 2.87 days, when its magnitude drops from 2.1 to 3.4; after about 10 hours it returns to normal. This star has widely been regarded as the most malevolent in the sky. In Greek myth it represents the evil eye in the severed head of the Gorgon Medusa, and its gaze turns all who see it into stone. In Hebrew astrology it represents either the "head of Satan", or Lilith, the first wife of Adam who became a nocturnal vampire.

NGC 86 (η Per), NGC 884 (χ Per) – The Double Cluster

Among several interesting deep-sky objects in Perseus, these open-star clusters, each around the size of the Full Moon, are visible to the naked eye, but magnificent through binoculars. Most of the stars are blue-white, with occasional reds. They are around 7,400 light years distant.

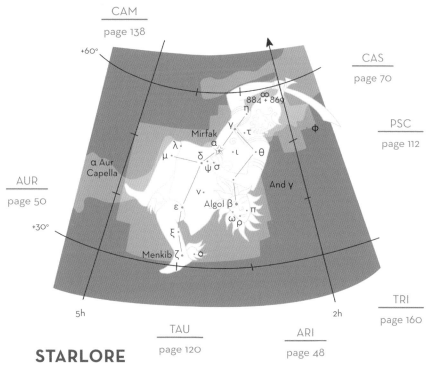

CAM
page 138

CAS
page 70

PSC
page 112

AUR
page 50

TRI
page 160

TAU
page 120

ARI
page 48

+60°

884 + 869

η γ τ

Mirfak
α ι θ
λ
μ δ σ φ
ψ σ

α Aur
Capella

ν

And γ

Algol β
ε π
ω ρ

+30°

ξ

Menkib ζ ο

5h   2h

STARLORE

The intrepid exploits of the hero Perseus began when he defended his mother Danaë from the attentions of King Polydectes. The king promised to find another bride if the young Perseus completed the seemingly impossible task of returning with the head of the Gorgon Medusa. She once had been an attendant of the goddess Athene, but was ravished in a temple by the god Poseidon (in Roman myth, Neptune), and for the loss of her chastity was transformed into a terrifying creature with snakes coiling from her head and a gaze that turned all mortals into stone. Before Perseus set out, the goddess Athene gave him a sickle to sever the Gorgon's head and a polished shield to see her reflection thereby avoiding her gaze. Perseus crept up on Medusa while she slept and struck off her head. On his return journey, he chanced upon the tragic scene of Andromeda (see pages 42–3).

Perseus holding aloft his sword, grasping the head of the Gorgon Medusa, from the Book of Fixed Stars by the Arabic astronomer al-Sufi. His star atlas was based on the astronomical studies of Ptolemy (2nd century BCE).

PSC – PISCIUM / THE FISHES

Pisces, the 12th zodiacal constellation, is difficult to discern because its stars are faint, none brighter than fourth magnitude. The figure consists of two fishes tied by a cord at their tails; the eastern fish swims vertically, in the general direction of the North Celestial Pole, while its companion strikes westward a few degrees above the equator and roughly parallel to it. A ring of five stars (ι, υ, γ, κ, λ), sometimes identified as the "Circlet", lies immediately south of the Great Square of Pegasus (see page 108), and south and slightly east of the bright star Markab (α Peg). The head of the north-swimming fish is about to buffet Andromeda (see page 42), and can be found just south of Mirach (β And). At the eastern edge of the constellation the cord that binds the fishes is marked by the star Alrischa.

Pisces culminates at midnight between late September and early October. It is widely visible, north or south of the equator, although in the Southern hemisphere, the figure begins to disappear for an observer at 57° South.

MAJOR STARS

α – Alrischa, 3.79, blue-white

Reflecting the symbolic duality of the two fishes, Pisces contains a number of double stars of interest to astronomers. Alrischa is a binary system, comprising stars of magnitudes 4.2 and 5.2, with an orbital period of 900 years. The system lies around 100 light years away. Its name, which may originally have come from the Babylonian word *riksu*, is Arabic for the "cord".

β – 4.53, blue-white

The westernmost main star in the second fish. Owing to precession (see pages 17–19) this will become a marker-star for the March equinox and the start of the Tropical zodiac, *c.*2813 CE.

η – 3.62, yellow

The constellation's lucida, by a fraction.

ω – 4.01, blue-white

The marker-star for the Aries equinox point of our era; it is just under 7° due north of the intersection of the ecliptic and the equator.

STARLORE

In Christian culture, Pisces has become identified with Christ as the "First Fish", who was born after the March equinox point had precessed from Aries back into Pisces,

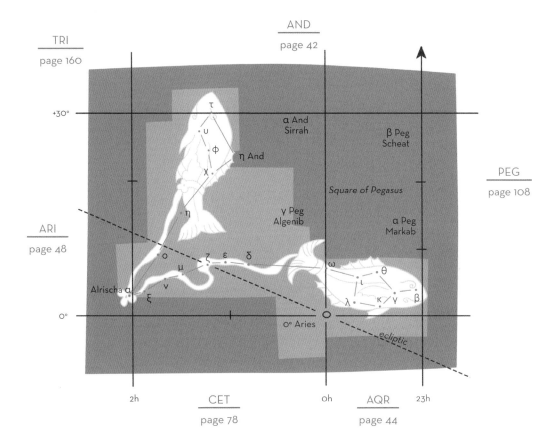

TRI
page 160

AND
page 42

PEG
page 108

ARI
page 48

CET
page 78

AQR
page 44

marking the transition to a new "Great Age" (see page 15).

The ancient form of Pisces is thought to have consisted of one fish. The Greek astronomer Eratosthenes (born 276 BCE) tells us that the origin of the fish symbolism is Derke, a Syrian goddess who was half-fish, half-woman.

The Romans compounded the idea of the fish-goddess in their myth of Venus and her son Cupid (in Greek myth, Aphrodite and Eros). They were startled by the monster Typhon, but Venus knew that they could escape by water. She grabbed Cupid and they plunged into the sea, where they were both transformed into fishes. To ensure that they did not lose each other they tied themselves together with a cord. In the heavens, we see mother and son, and the cord that binds them in love.

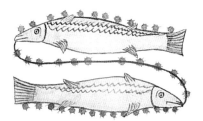

In many representations, such as this from the 13th century, the fishes of Pisces are shown linked by their mouths rather than their tails.

PSA — PISCIS AUSTRINI / THE SOUTHERN FISH

Piscis Austrinus (sometimes termed Piscis Australis) lies approximately 30° South of the equator; its designation as "Southern" distinguishes it from the Northern fishes of Pisces. Although moderately sized and (with one exception) made up of dim stars, it is easily identified from its location at the feet of the water-carrier Aquarius. It largely comprises stars of the fourth and fifth magnitudes. Its exception is the bright star Fomalhaut, marking the mouth of the fish and located south of the spout of the water-carrier's jug (see Aquarius, pages 44–5). The figure appears to be swimming upstream along the "River of Aquarius", which pours in a curve from the jug — the fish is often thought to be consuming this stream in its great mouth. It is commonly depicted with its back facing north, but in some old star atlases it appears instead with its belly upward. Piscis Austrinus culminates at midnight in late August. It is a fine constellation for the observer in the Tropics and the Southern hemisphere, but anyone at middle-latitudes North sees only a faint figure, which clings low on the horizon on light summer evenings. Upward from 53° North the constellation progressively disappears from view, although under ideal conditions Fomalhaut can still be glimpsed on the horizon as far north as 60°.

MAJOR STARS

α — Fomalhaut, 1.16, blue-white
This is the 18th brightest star in the sky, and traditionally a major navigational star. Its name comes from the Arabic for "fish's mouth". It lies 22 light years distant.

STARLORE

The Southern fish was well known to the early Greeks very much in the form we see today. However, the mythology of the figure as a whole has often been subsumed into the story of its famous lucida, Fomalhaut. Although it falls well south of the ecliptic, Fomalhaut's status as the brightest star in this region of the sky made it a valuable marker of the seasons, and it was accordingly one of the four Royal Stars or heavenly "watchers" identified in ancient Mesopotamia — the others being Regulus (α Leo), Aldebaran (α Tau) and Antares (α Sco). The figure as a whole, and Fomalhaut in particular, have frequently been identified with the zodiacal constellation Aquarius, thus completing a celestial cross of the four signs, Taurus, Leo, Scorpio and Aquarius.

In his book *The Patterns in the Sky* (1988), the astronomer Julius Staal (1917–1986) traces Piscis Austrinus back to Egyptian mythology.

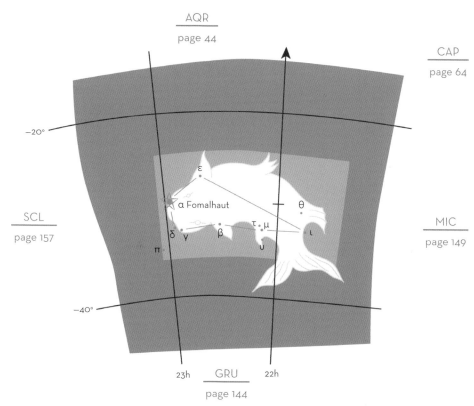

AQR
page 44

CAP
page 64

SCL
page 157

MIC
page 149

GRU
page 144

−20°

−40°

23h   22h

ε

α Fomalhaut   θ

δ γ   β   τ · μ   ι
υ
π

The god-king Osiris, who was said to have brought civilization to Egypt, was murdered by his jealous brother Set, who cut up the body into 14 parts, which were then cast into the Nile. Isis, the sister and consort of Osiris, searched for the body-parts, finding them all except the phallus. This had been swallowed by the Nile crab, Oxyrhynchus shown mirrored by the constellation Piscis Austrinus swallowing the waters of life, .

In flood myths, this act of swallowing the water flowing from Aquarius' jug has been seen as salvation from the deluge. Piscis Austrinus has also been identified as the parent of the fish of Pisces.

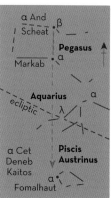

α And
Scheat   β

Pegasus
α
Markab

Aquarius   α
ecliptic   λ

α Cet   Piscis
Deneb   Austrinus
Kaitos   α
Fomalhaut

To find Piscis Austrinus, locate the Square of Pegasus (see page 109) and extend the line made by the right-hand edge of the square (α and β Peg), to cross over Aquarius, and lead directly to bright Fomalhaut (α PsA).

SAGITTARIUS

SGR — SAGITTARII / THE ARCHER

The ninth zodiacal constellation, Sagittarius is a centaur, half-man, half-horse, aiming an arrow toward Scorpius. Prominent in the Southern hemisphere, it culminates at midnight in June and July. Above middle latitudes in the North, it sits low on the southern horizon, partly obscured. It is on the eastern side of the Milky Way, 25° southwest of Altair (α Aql; pages 46–7). If the archer were to raise his arrow a few degrees, his aim would be through the Milky Way to the Galactic Centre. The Sun appears in this constellation at the December solstice.

MAJOR STARS

α — Rukbat, 4.1, blue-white
The "knee". Despite being α, this is not the brightest star in the constellation (see ε below).

β1 and β2 — Arkab Prior and Arkab Posterior, 4.3 and 4.5, blue-white and white
An unrelated but visually close pair. "Arkab" is the Achilles tendon from the calf to the heel.

γ — Alnasl, 3.0, yellow
This star marks the "point" of the arrow.

ε — Kaus Australis, 1.9, blue-white
The "southern bow" is the lucida, a giant, 88 light years distant. It marks the archer's bow, together with the third-magnitude Kaus Meridionalis (δ Sgr) and Kaus Borealis (λ Sgr).

σ — Nunki, 2.0, blue-white
Nunki marks the hand of the archer, drawing back the arrow. In Assyro-Babylonian times, it was known as "the star proclaiming the sea". The "sea" is the portion of sky that rises to the east after Sagittarius and contains Aquarius, Capricornus, Delphinus, Cetus, Pisces and Piscis Austrinus, all associated with water.

ζ — Ascella, 2.6
The name for this star derives from the Latin *axilla*, meaning "armpit".

M8, M17, M22
Within or at the borders of naked-eye observation are M8, the Lagoon Nebula — a large patch equivalent in area to three Full Moons — and the globular cluster M22, visible at fifth magnitude but best observed through binoculars. M17 is the Omega or Horseshoe Nebula.

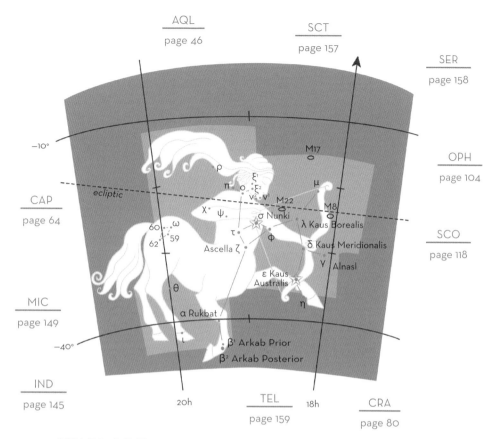

AQL
page 46

SCT
page 157

SER
page 158

OPH
page 104

CAP
page 64

SCO
page 118

MIC
page 149

IND
page 145

TEL
page 159

CRA
page 80

−10°

ecliptic

M17

ρ

ξ¹
π   ξ²
ν² ν¹   M22

μ

M8

χ ψ

σ Nunki

λ Kaus Borealis

τ

φ

δ Kaus Meridionalis

60 ω
62   59

Ascella ζ

γ Alnasl

θ

ε Kaus
Australis

η

α Rukbat

−40°

ι

β¹ Arkab Prior
β² Arkab Posterior

20h

18h

STARLORE

The Greeks associated Sagittarius with Crotus, the satyr (part man, part goat, with a long, horse-like tail), and it was often illustrated standing on two legs. The Roman mythographers identified him with the gentle and wise centaur Chiron, which has led to frequent confusion with the Southern constellation Centaurus (see pages 74–5). However, there is a distinct difference between the two creatures: Sagittarius is a hunter, traceable to the Mesopotamian archer-god Nergal, who was associated with the wrathful god Irra of war and fire (in Greek myth, Ares; in Roman, Mars).

Sagittarius the Archer, from an Arabic manuscript. The figure is oriented as if viewed from the outside of the Celestial Sphere (that is, he shoots his arrow from right to left, rather than left to right).

SCORPIUS

SCO – SCORPII / THE SCORPION

Commonly known as Scorpio, Scorpius is the eighth zodiacal constellation. It lies across the Milky Way, with Ophiuchus to the north, and Lupus and Ara to the south. The body stretches well south of the ecliptic, making this a magnificent constellation for equatorial and Southern observation, culminating at midnight in June. It loses impact above mid-latitudes North, where the distinct hook of its tail and sting becomes faint and indiscernible at the horizon.

MAJOR STARS

α – Antares, averaging 1.35, red
A supergiant 400 times the diameter of the Sun, 170 light years distant. Its brightness fluctuates over a 4.75-year cycle. At latitude 5° South, it is an ecliptic marker, one of the four Royal Stars (heavenly "watchers") of Mesopotamia, forming a great cross on or near the ecliptic; the other stars are Aldebaran (α Tau), Regulus (α Leo) and Fomalhaut (α PsA). Antares means "rival to Ares" or "equivalent to Ares" — Ares being the Greek form of Roman Mars, the red planet. The star is occasionally *Cor Scorpii*, "heart of the scorpion".

β – Acrab or Graffias, 2.6, blue-white
"Scorpion" and "claws". Confusion occurs in the star-lists because the latter name has also been applied, somewhat illogically, to ζ Sco. The star, a double, lies just above the ecliptic.

δ – Dschubba, 2.3, blue-white
The "front" or "forehead" of the scorpion.

λ – Shaula, 1.6, blue-white
The "sting" lying just beside υ Sco, Lesath (2.7), a name that derives from another word for the sting, *Al Las'ah*.

θ – Sargas, 1.9, yellow
A giant at 190 light years distance. The name is of Mesopotamian origin.

In Maori tradition Scorpius represents the fish-hook, shown by these carved jade beads, of the ancestral hero Maui. One day, while fishing, Maui caught and pulled up from the ocean a piece of land. Gradually, the edges of this land became serrated, so much so that the land broke in two. This is how New Zealand came into existence. The hook dislodged from the island with such force that it flew into the sky, where it has remained ever since.

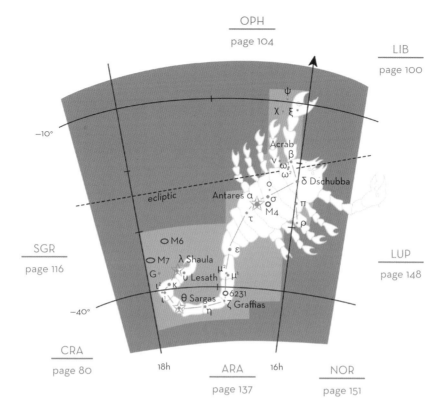

OPH
page 104

LIB
page 100

SGR
page 116

LUP
page 148

CRA
page 80

ARA
page 137

NOR
page 151

STARLORE

The lucida of this ancient constellation, Antares (α Sco), marked the place of the Sun at the September equinox at the beginning of Mesopotamian civilization, some five millennia ago. In Egypt the stars of Scorpius were, for a period, seen as a serpent. The figure was once much larger: in the classical Greek and Roman periods of the first centuries BCE, the scorpion's huge claws included the stars that now form Libra (see page 100).

Scorpius has always suffered a baneful reputation. In some accounts of Greek myth, the Earth-goddess Gaea commanded the scorpion to sting Orion, who was then revived by Aesculapius (see Ophiuchus; pages 104–5).

In the turning sky, as Scorpius rises in the east, Orion dies in the west. The subsequent rising of Orion is his restoration. As he appears in the east, Scorpius sets, crushed by Ophiuchus.

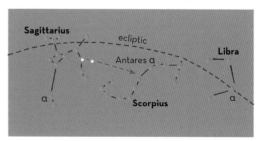

The line made by the arrow of Sagittarius points directly into the body of the scorpion, westward along the ecliptic. As an additional ecliptic marker, use Libra's α-star.

TAURUS

TAU – TAURI / THE BULL

With its rich array of stars including its famous clusters of the Pleiades and Hyades, the second zodiacal constellation forms one of the most impressive figures of the Northern sky. Taurus lies northwest of the giant Orion, and southwest of Auriga; it is immediately identifiable from its lucida Aldebaran, the red eye of the bull, close by the loose group of the Hyades, which mark the face. Our conventional image for Taurus is an incomplete figure, depicting the front end of a bull, which faces east, with its head, which has greatly exaggerated horns, bowed toward Orion. The tip of the northern horn touches the heel of Auriga, and the star here, Elnath, was formerly shared between both constellations.

The location of Taurus close to the celestial equator ensures that, apart from some obscuration in the Antarctic region, the main part of the figure is visible worldwide. It appears at its most magnificent from the Tropics and in the Northern hemisphere's winter skies, culminating at midnight in late November and early December.

MAJOR STARS

α – Aldebaran, 0.85, pale red

An irregular variable at a distance of 68 light years. Aldebaran means "the follower", because it follows either the Pleiades or, more probably, the Hyades, rising just after they have risen and setting after they have set. Lying within 6° of the ecliptic, this star was one of the four Royal stars or "watchers" of ancient Mesopotamia; the other three are Regulus (α Leo), Antares (α Sco) and Fomalhaut (α PsA).

β – Elnath, 1.65, blue-white

Elnath, sometimes Al Nath, is Arabic for "the butting one". This star was once the γ-star of Auriga; it is now astronomically assigned to Taurus.

M1 – The Crab Nebula

This deep-sky object may just be discerned using binoculars, approximately 1° northwest of ζ Tau on the southern horn. It is the remnant of a famous supernova explosion observed in 1054 CE; its distance is around 6,500 light years. It gained its name because its extended filaments resemble the pincers of a crab.

STARLORE

The half-bull motif has been found in a Babylonian record of c.2000 BCE. Whether this was associated with Taurus remains unproven. But certainly these stars were venerated as marking the place of the Sun at the March equinox some five millennia ago.

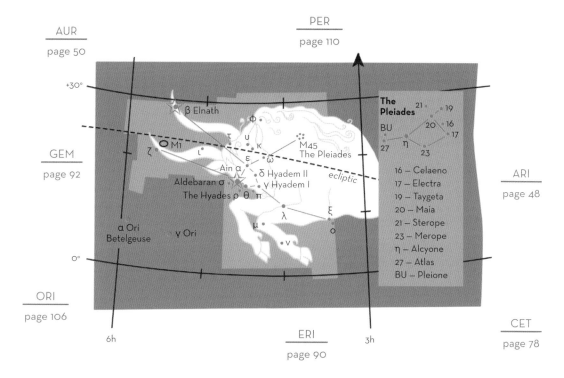

AUR
page 50

PER
page 110

GEM
page 92

ARI
page 48

ORI
page 106

ERI
page 90

CET
page 78

+30°

β Elnath

Φ

τ υ
κ
ζ ○ M1
ι ε
Ain α ω
δ Hyadem II
Aldebaran σ γ Hyadem I
The Hyades ρ θ π
λ
ξ
μ ο
ν

M45
The Pleiades

ecliptic

α Ori
Betelgeuse
γ Ori

0°

6h
3h

The Pleiades
BU

21 19
20 16
17
27 η 23

16 — Celaeno
17 — Electra
19 — Taygeta
20 — Maia
21 — Sterope
23 — Merope
η — Alcyone
27 — Atlas
BU — Pleione

The symbolism of the bull or cow occurs worldwide for Taurus. Since late-Egyptian culture in the last centuries BCE, Osiris, shown as a bull-god, was identified with the constellation, as was his cow-goddess sister Isis. The crescent Moon formed her horns, which may be the origin of Taurus' astrological sigil (see page 18).

The Greek myths give us two tales of the lusts of Zeus: Io, turned into a cow by Hera (see Pavo, page 152); and Europa, seduced by Zeus as a gentle, white bull on the shore. The moment Europa climbed upon the bull's back, it carried her across the ocean to Crete, where Zeus ravished her.

The Persian solar cult of the bull-god Mithras became widespread in the Roman Empire. The Romans saw the wine-god Bacchus in Taurus. During Bacchic festivals a bull, decked with flowers, was escorted by dancing girls, who were represented by the Hyades and the Pleiades.

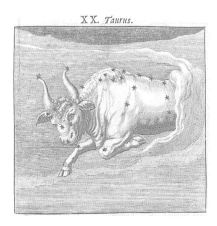

XX. Taurus.

Taurus, typically shown without its hind quarters in a star atlas of 1681.

THE PLEIADES

This most famous of all star clusters lies to the northwest of Aldebaran, on the shoulder-blade of the mighty bull. It is a useful marker for the Tropic of Cancer, which it approaches to within a degree.

Although these stars are universally known as the "seven sisters", there are actually between six and eight stars, or just possibly nine, that are discernible to the naked eye, led by the brightest of the group, Alcyone.

Binoculars reveal several dozens of the hundreds of stars in the cluster, and a larger nebulous area spanning three diameters of the Moon. The cluster is 410 light years distant; its stars, for the past 50 million years, have been forming out of a cloud of stellar dust.

The Pleiades have excited interest since antiquity, and have occasionally been treated as a constellation in their own right. For the Hindus they were a flame dedicated to the fire-god Agni, or sometimes a short-handled razor. They have frequently been seen as

birds, and were popularly known as "hen and chickens" in medieval Europe. In reference to Taurus' association with Bacchus, the god of wine and revelry, the Pleiades have been seen as a bunch of grapes.

However, the most enduring and widespread tradition sees this cluster as seven girls or sisters. For the Greeks they were the daughters of Atlas and Pleione, who figure as the eighth and ninth stars of the group. Listed in descending order of brightness, the nine stars including the parents are: Alcyone, Atlas, Electra, Maia, Merope, Taygeta, Pleione, Celaeno and Sterope. It is an interesting test of observation to see how far down the list a star-watcher can get: spotting Alcyone (magnitude 2.9) is an easy task; Sterope (magnitude 5.8) is usually beyond the range of the naked eye.

According to one legend the Pleiades were the virgin consorts of the goddess Artemis. When they were pursued by the hunter Orion, the gods heard their appeal for help and placed them in the sky as doves. Like the Hyades, the Pleiades spend their time weeping, and there are several different reasons given for their distress, quite apart from the unpleasantness of being molested by Orion. One suggestion is that the sisters weep for a lost companion, possibly a star that has dimmed considerably in historical times; this could be Sterope, who is so faint as to be easily lost.

An Italian manuscript illustration (9th–10th century CE) of the seven sisters of the Pleiades, who were the daughters of the Titan Atlas.

THE HYADES

The Hyades is a beautiful cluster of stars, the brightest of which form a distinct v-shape in the face of the bull. Aldebaran (α Tau) lies on the eastern edge of the cluster, but does not belong to it. The whole group is a fine sight in binoculars, covering more than 5° of the sky; it comprises some 200 individual stars, lying at a distance of 150 light years. The brightest star in the cluster is θ² Tau (3.4).

In the Greek myths the Hyades have been individually named as follows: Aesula, Ambrosia, Dione, Thyene, Koronis, Eudora and Polyxo; but none of these names is attached to any particular star in the cluster.

Hyades means "rainy ones". The stars were of ill omen to farmers and sailors because the season of storms and heavy rain coincided with the time of their heliacal rising and setting (their first appearance after a time of invisibility, and their last appearance before becoming invisible again); during the classical period these times were at the end of May and November. The Roman poet Ovid (43 BCE—17 CE) records that the sisters were grief-stricken when their mortal brother Hyas drowned in a well, and their tears fall on us as rain.

One Roman tradition identified the sisters rather unflatteringly as "little pigs", derived from a variant etymological interpretation of the ancient Greek name. The poet Pliny (1st century CE) made a curious attempt at rationalizing conflicting strands of interpretation by suggesting that the continuous rains associated with these stars made the roads so muddy that the sisters appeared to wallow in them, like pigs!

Arabic authors occasionally interpreted the Hyades as "little she-camels", with the bright star Aldebaran representing the Large Camel.

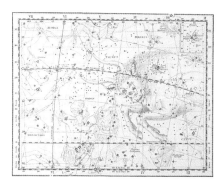

This illustration from the Whittaker Star Atlas of 1822 shows Taurus, surrounded by its neighbouring constellations. In the bull's face lie the stars of the Hyades, just next to the bright star Aldebaran, the red eye of the bull. The Hyades were the daughters of the Titan Atlas and Aethra, and so were half-sisters to the Pleiades.

UMA – URSAE MAJORIS / THE GREATER BEAR

Ursa Major, the third largest constellation, covers a sizeable chunk of the circumpolar Northern sky. However, the figure is insignificant compared with the world-famous asterism of seven stars that forms the rump and tail of the bear, known as the Plough or Big Dipper. The pattern of this asterism is so distinctive that it is an ideal starting-point from which to orient the sky. Further, it includes two stars (α and β UMa) on the far side of the Plough or Big Dipper away from the "handle" that, when joined, form a line to the North Pole. From beyond 40° South the asterism cannot be fully observed, and Ursa Major as a whole is lost at middle latitudes of the Southern hemisphere. Its midnight culmination is in March. (See also page 12.)

MAJOR STARS

α – Dubhe, 1.8, yellow
The name derives from the Arabic for "bear". Despite being α this is not the lucida: see ε.

β – Merak, 2.4, greenish white
This star is so called because it marks the "flank" (of the bear).

γ – Phecda, 2.4, yellow-white
The name means "thigh".

δ – Megrez, 3.4, white
The faintest of the stars of the Plough or Big Dipper, its name means "root of the tail".

ε – Alioth, 1.8, white
The name is of uncertain derivation. This is the constellation's lucida.

η – Alkaid or Benetnash, 1.9, brilliant white
The "principal mourner" (from a derivation of the Arabic *Ka'id Banat al Na'ash*) of the children of Al Na'ash, who were murdered by the Pole Star, Al Jadi, according to Arabic myth. Every night as the stars of the Plough or Big Dipper they still crowd around in their circumpolar course, seeking revenge.

ζ – Mizar, 2.4, white
The name has ambiguous origins. The Arabs associated this star with the mourning of Alkaid. It has a well-known fourth-magnitude companion, Alcor, "the rider".

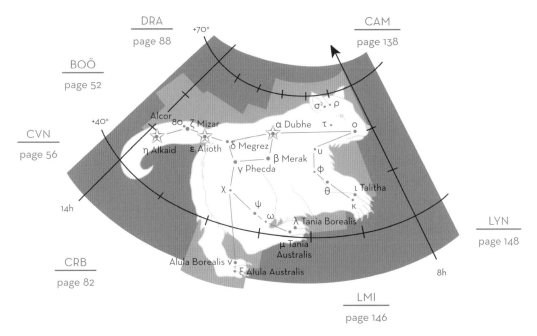

DRA
page 88

CAM
page 138

BOÖ
page 52

CVN
page 56

LYN
page 148

CRB
page 82

LMI
page 146

STARLORE

From early times Ursa Major has been paired with Ursa Minor. In one account, every year the god Cronus swallowed the children borne of his wife Rhea. But one year instead of giving him the baby Zeus (Jupiter in Roman myth), she gave him a stone wrapped in baby's clothes, while Zeus was hidden, and nursed by the nymphs Helice and Cynosura. Cronus pursued the infant, but Zeus slipped away. Before he escaped he placed his nurses in the heavens: Helice as the Greater Bear, Cynosura as the Lesser.

Alternatively, the nymph Callisto, a servant of the huntress Artemis (Diana), was raped by Zeus and made pregnant, causing Artemis to banish her for impurity. From the coupling she bore the child Arcas. In a fit of jealousy, Zeus' wife Hera turned Callisto into a bear, which slunk away into the forest. Arcas grew up to become a hunter. One day while he was hunting, Callisto heard his voice and rushed to greet him. Arcas was about to kill her when Zeus intervened and sent both mother and son into the heavens as the Greater and Lesser Bears.

This Chinese print showing Ursa Major draws out not only the Great Bear, but also the seven stars of the Plough or Big Dipper as a chariot in the bear's tail.

THE NORTH POLE

SIGNPOST MAP 2

Shown here are the Northern circumpolar constellations. The seven stars in Ursa Major that make up the widely-known Plough or Big Dipper are picked out in white. The "pointers" to the pole are β and α UMa, Merak and Dubhe. Cepheus and Cassiopeia are useful markers for the *equinoctial colure*, the circle that runs through both poles and the two equinox points in Aries and Virgo. The tail of Ursa Minor (the constellation that mimics on a small scale the shape of the Plough or Big Dipper) marks the *solsticial colure*, which runs through both poles and the solstice points in Cancer and Capricornus. The view shown here is for midnight local mean time (1am with Daylight Saving) at the solstice, June 22.

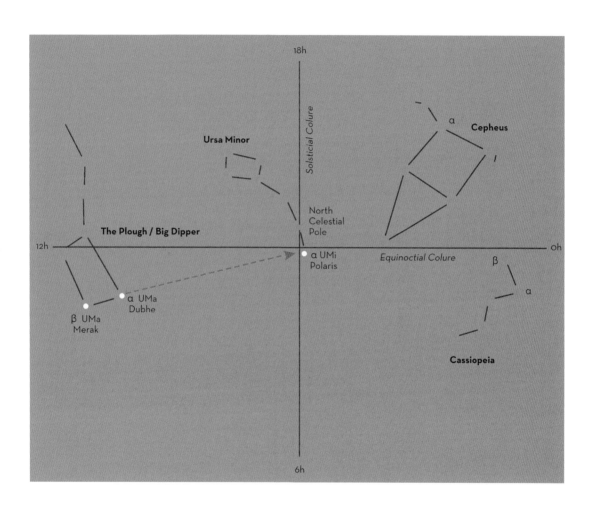

18h

Solsticial Colure

Ursa Minor

α **Cepheus**

North
Celestial
Pole

The Plough / Big Dipper

12h

α UMi
Polaris

Equinoctial Colure

0h

β

α UMa
Dubhe

α

β UMa
Merak

Cassiopeia

6h

URSA MINOR

UMI – URSAE MINORIS / THE LITTLE BEAR

This constellation came into practical usage in classical times through the astronomer Thales (*c*.600 BCE), who noted that the Phoenician mariners of his day used these circumpolar stars for navigation in preference to Ursa Major.

The group forms a pattern of seven stars like a reversed Plough or Big Dipper. In our epoch, the last star in Ursa Minor's tail very closely marks the Celestial North Pole; hence we call it Polaris (α UMi), the Pole Star.

MAJOR STARS

α – Polaris, 2.0, yellow

A supergiant. The lucida's place of honour has been acknowledged under various names in many cultures. It was worshipped in early Hinduism as the god Dhruva, and called "pivot of the planets". The Arabs termed it *Al Kutb*, "axle". It was also, more curiously, *Giedi*, the slayer of the man whose mourners cluster in Ursa Major (see η UMa, page 124).

β – Kochab, 2.1, orange

A giant star only a fraction less bright than α. Approximately 3,000 years ago, the North Celestial Pole would have been much closer to this star than to Polaris.

STARLORE

Most myths treat this constellation together with Ursa Major. However, the German cosmographer Petrus Apianus (1495–1522) attributed a mythical connection of their own to these stars. He believed they were the Hesperides, the seven nymph daughters of the Titan Atlas. Their names were Arethusa, Aegle, Erythea, Hestia, Hespera, Hesperusa and Hespereia. In the far west of the world, on Mount Atlas, they tended the tree of golden apples later given to Hera by the Earth-goddess Gaea as a wedding gift on her marriage to the supreme god Zeus (in Roman myth, Jupiter).

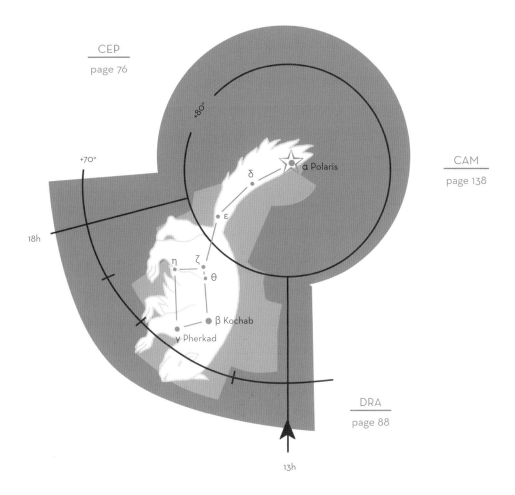

CEP
page 76

CAM
page 138

+80°

+70°

18h

δ

ε

η ζ
θ

β Kochab
γ Pherkad

☆ α Polaris

DRA
page 88

13h

THE MAJOR CONSTELLATIONS

VIR – VIRGINIS / THE VIRGIN

The sixth zodiacal constellation, Virgo is second only to Hydra in area. However, apart from its magnificent lucida Spica, it presents an indistinct grouping to the casual observer. The figure of a girl, usually with wings, straddles the equator, mostly to the north of the ecliptic, although Spica, an ecliptic marker, falls just 2° south of the circle. This brilliant star is easily found in the Northern hemisphere in spring and early summer evenings by taking a line through the handle of the Plough or Big Dipper (see pages 124–5) and extending the curve southward through Arcturus (α Boö) down to Spica (see pointer map, page 53). From the Southern hemisphere, Virgo is an autumn constellation located 30–40° north of Centaurus. Spica lies roughly midway on a vast 100° arc of the ecliptic stretching between two other first-magnitude ecliptic markers: Antares (α Sco) and Regulus (α Leo). In our era, the September equinox point falls near β Vir.

MAJOR STARS

α – Spica, 1.0, blue or blue-white
The Virgin's "spike" marks the ear of wheat in Virgo's left hand. It is around 260 light years away. The desert Arabs occasionally named this star Azimech, from Al Simak, the "defenceless" or "unarmed one", unattended by any nearby stars.

β – Zavijava, 3.8, pale yellow
The name derives from the Arabic for "corner": in early times this star marked the corner of a kennel for dogs that barked at the heels of nearby Leo.

γ – Porrima, 2.8, yellowish-white
An alternative name for Carmenta, the Roman goddess of prophecy who gave inspiration to poets. This is a pair of stars, both magnitude 3.5, orbiting each other every 169 years.

ε – Vindemiatrix, 2.8, yellow
This star marks the right arm of the Virgin, in which she cradles a palm frond. The name is from the Latin for "female grape-gatherer", as in early times its heliacal rising marked the time of wine-making. In astrology this is thought to be an unfortunate star. Vindemiatrix lies at a distance of 100 light years.

STARLORE

Major elements of the description of Virgo date as far back as Assyro-Babylonian culture. The constellation has aways been considered female and has been especially associated with the tension between fertility and purity — strands that paradoxically entwine in her myths.

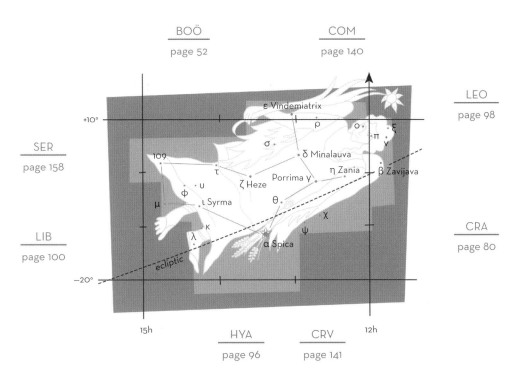

BOÖ
page 52

COM
page 140

LEO
page 98

SER
page 158

LIB
page 100

CRA
page 80

HYA
page 96

CRV
page 141

ε Vindemiatrix
ρ
ο
π
ξ
σ
δ Minalauva
109
τ
η Zania
β Zavijava
φ
υ
ζ Heze
Porrima γ
μ
ι Syrma
θ
χ
κ
ψ
λ
α Spica

+10°

−20°

ecliptic

15h

12h

The Babylonians linked her with the goddess Ishtar, also known as Ashtoreth or Astarte. The latter is the forerunner of the Saxon fertility and spring-goddess Eostre, whose festival is the origin of Easter, at a time of year when the constellation Virgo is becoming prominent in the evening skies.

One myth about Ishtar recounts that she descended into the underworld to recover her dead lover, the harvest-god Tammuz. She became imprisoned here, and in her desolation brought blight upon the world, forcing the great gods to release her. This parallels the Greek myth of the beautiful Persephone (in Roman myth, Proserpina), who was abducted by Hades (Pluto) and taken to the underworld, causing her mother Demeter (Ceres) to destroy the harvests.

A 5th–3rd-century-BCE amulet showing the Egyptian goddess Isis, associated with the constellation Virgo. Spica is the sheaf of corn dropped by Isis as she fled from a monster.

THE MAJOR CONSTELLATIONS

On a clear night during the darkest part of each month, when the Moon is near to New, we may see in all its beauty the ribbon of the Milky Way. We will then be gazing along the plane of the flattened disk of an island of stars, our home Galaxy. From our perspective the part of the Galaxy that we instantly recognize as the Milky Way (the most densely populated part, holding nine tenths of all visible stars) covers a mere one tenth of the visible heavens. The Galaxy is approximately 100,000 light years in diameter and 2,000 light years thick.

The Sun is one of an estimated 100,000 million stars in the whole Galaxy. It lies on one of the spiral arms approximately two thirds of the way out from the Galactic Centre, the hub of the system in the rich star-fields of Sagittarius.

The "Magellanic Clouds", which appear to be broken-off parts of the Milky Way, are two small companion galaxies associated with our system (see pages 142 and 164). There are countless other galaxies too, at enormous distances from our own. Ours is the second largest of a cluster of around 30 that make up our "Local Group".

The Milky Way has been a focus of imaginative inspiration from earliest times. Almost universally it has been seen as a heavenly river or road. In the Hebrew tradition it was the River of Light; in India it was the reflection of the earthly River Ganges; in ancient Egypt it was the celestial counterpart of the River Nile. Another frequently found motif is that of the Milky Way as the pathway of souls. In this view, the gateways between heaven and earth are at the intersections of the Milky Way and the ecliptic, in the constellations Sagittarius and Gemini.

Greek myth has a colourful account of the formation of the Milky Way. In order to win

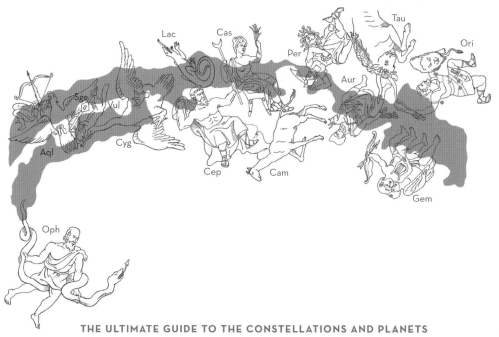

the hand of Alcmene, the youth Amphitryon was required to avenge the death of Alcmene's brothers. On the very night that Amphitryon fulfilled his pledge, the supreme god Zeus (in Roman myth, Jupiter) disguised himself as the youth and stole into Alcmene's chamber. He assured her that the act of vengeance had been fulfilled, and lay with her. The result of their union was the hero Heracles (Hercules; see pages 94–5).

Zeus' wife Hera was notoriously jealous of her husband's infidelities, usually taking vengeance on either the rival, or the children of Zeus' extra-marital couplings. On this occasion, however, Zeus outwitted Hera. He arranged for her to come across the infant as if abandoned, and in compassion the goddess began to suckle it, thereby granting Heracles the gift of immortality through her milk. But Heracles gave Hera's nipple such a lusty tug that the goddess screamed in pain. As she pulled the child away from her, a fountain gushed from her breast. Some of the milk fell onto the earth, to form lilies, but most of it shot into the sky, to become the Milky Way.

In the ancient tradition of the village Misminay in Peru, the Milky Way is thought to take water from the cosmic ocean in which the Earth floats, which it then sends back to us as rain. The local people believe that their River Vilcanota is an earthly mirroring of the celestial river. Furthermore, the dark "shapes" in the Milky Way, caused by patches of interstellar dust, collectively called by the Andeans the *Pachatira*, are given names (for example, Baby Llama, Toad and Serpent), as we give names to the constellations.

The Milky Way passes through a jumble of constellations of the Northern (below) and Southern (opposite) skies. The dots are those of the 50 brightest stars in the sky that fall on this celestial river. (For key to abbreviations, see page 27.)

A star map of 1660 largely showing the Southern sky. The constellations are superimposed on terrestrial continents: South America can be found below Cetus' tail; beneath the River Eridanus is Africa; Tucana sits on the Antarctic.

A 13th-century stone carving from an arch showing Piscis Austrinus (the Southern Fish) over Ara (the Altar) — although these constellations are actually far away from one another. In Greek lore Ara represents the altar of sacrifice to Zeus.

THE MINOR CONSTELLATIONS

" Over all the sky — the sky! far, far out of reach,
studded, breaking out, the eternal stars. "

Walt Whitman (1819–1892), Bivouac on a Mountain Side

Our survey of the fixed stars concludes with the remaining 48 constellations. A few of these are ancient and have a rich mythology, but most are relatively modern. Some modern constellations, such as Camelopardalis, are found in the Tropical and Northern skies, which have been mapped by Northern-hemisphere civilizations since ancient times. However, the main group comprises Southern-hemisphere stars mapped by European voyagers from the 15th century onward, sometimes containing bright but nameless stars. The Greeks called stars lying outside their figures "unformed" or "dispersed", but astronomy's need to fill in the spaces has led to new groupings. Fitting uncomfortably in a cosmos teeming with archaic myth, they are often little more than oddments of space between ancient and established shapes. Nevertheless, some represent the honourable recognition of historical figures and inventions, while others show exotic animals. Many, too, provide us with a fascinating observational challenge.

ANTLIA

ANT – ANTLIAE / THE AIR PUMP

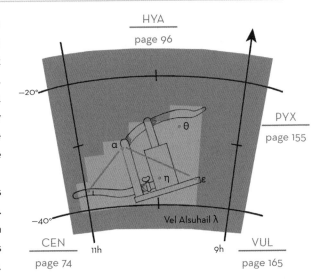

Antlia is an obscure grouping that lies to the north of Vela, the sail. It was created in the 18th century by Abbé Nicolas de Lacaille, the first cosmographer fully to map the Southern skies. The industrious Abbé was responsible for 14 of the new constellations, most of them equally inconspicuous; here he wished to commemorate the air pump, which had been invented by the physicist Denis Papin (1647–c.1712).

None of Antlia's stars are named, but this is not unusual among the Minor Constellations. The α star, Antlia's brightest, lies only at the dim end of the 4th magnitude. The constellation's midnight culmination is at the end of February.

APUS

APS – APODIS / THE BIRD OF PARADISE

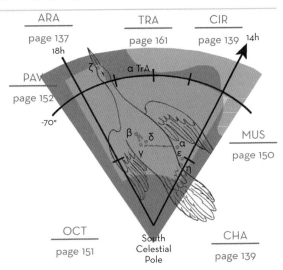

First given the name *Avis Indica*, the Indian Bird, in the 1590s, Apus was mapped by the Dutch navigators Pieter Keyser and Frederick de Houtman when they voyaged to the South. It may be located easily from Triangulum Australe, as it lies between the base stars of the triangle, α and γ, and the South Celestial Pole. The main stars are located approximately 13° from the pole.

The α star is an orange giant, 230 light years distant. β and γ are also orange. δ is a pair of 5th-magnitude orange stars.

The Victorian cosmographer R.H. Allen, in his *Star Names*, records that in China this asterism was named "Curious Sparrow".

ARA – ARAE / THE ALTAR

Ara lies on a dense part of the Milky Way immediately south of the hook of stars making up the tail of Scorpius. Its midnight culmination is around June 10. Best seen from the Tropics and the Southern hemisphere, it is invisible from middle latitudes in the North.

Although Ara is a minor and relatively indistinct constellation with no named stars, it has been known since ancient times. It was usually shown as an altar on which incense is burnt; and occasionally it was envisaged as a pyre placed on the summit of a temple or a tower, or serving as a lighthouse. To the poets of Greece and Rome, Ara represents the heavenly altar created by the gods of Mount Olympus to celebrate their defeat of the Titans and to consecrate their new-found status. At this highest altar the gods swore their allegiance to the supreme god Zeus (in Roman myth, Jupiter). The smoke from the altar-fire was said to pour out to create the Milky Way.

Ara has several stars of the 3rd magnitude: α (blue-white); β (the lucida; orange); γ (a blue supergiant, 1,800 years away); and ζ (orange). NGC 6193 is a 5th-magnitude star cluster certainly worth a view through binoculars. It consists of about 30 stars, which are 4,400 light years away. The brightest star in the group is magnitude 5.7 and blue-white in colour.

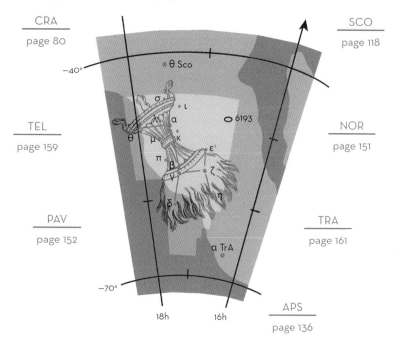

−40°

θ Sco

σ

ι

λ α

θ

μ κ

π

β

γ

ε¹

6193

ζ

η

δ

α TrA

−70°

18h

16h

CAELUM

CAE – CAELI / THE CHISEL

This obscure constellation, one of 14 created by Abbé Nicolas de Lacaille in his investigation of the Southern sky (1751–2), lies northwest of Canopus (α Car; page 66), and between the southeast shores of Eridanus and the dove Columba. Caelum falls from view at middle-Northern latitudes. It culminates at midnight in early December.

Caelum was for a time known as *Scalptorium*, another term for a sculptor's tool. In the 19th century the American astronomer Elijah Burritt tried to rename the constellation as *Praxiteles*, after the Greek sculptor of the 4th century BCE – no doubt an attempt to bring de Lacaille's creation back into line with classical motifs.

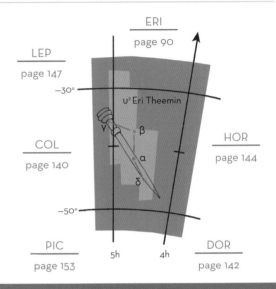

CAMELOPARDALIS

CAM – CAMELOPARDALIS / THE GIRAFFE

Camelopardalis (or Camelopardus), a faint Northern constellation, lies mostly north of Auriga and Perseus. The α and β stars roughly form a north-south line, extending to Capella (α Aur). The pair culminates at midnight around December 6. In the Southern hemisphere it fades from view at middle latitudes.

First published by the Dutch theologian Petrus Plancius in 1613, it represents the camel that carried Rebecca into Canaan to marry Isaac in the Biblical story.

The α star (a blue supergiant; 4.29) lies at more than 3,000 light years; its absolute magnitude is thought to be an amazing –6.

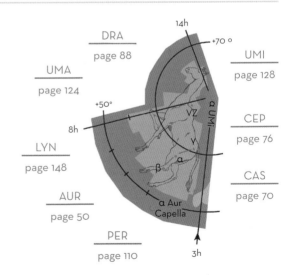

CHAMAELEON

CHA – CHAMAELEONTIS / THE CHAMELEON

This small Southern circumpolar constellation was introduced by the Dutch navigators Pieter Keyser and Frederick de Houtman in their travels of 1595–7. It lies south of Miaplacidus in Carina, and borders the polar constellation Octans. It culminates at midnight at the beginning of March.

The explorers of the 15th and 16th centuries filled up the Southern skies with diverse figures, which often represented such newly-encountered birds or animals.

The constellation's lucida is the α star (magnitude 4.07, white). δ is made up of a pair of unrelated stars that are easily split by binoculars: δ² (magnitude 4.4) is a blue star, at 780 light years.

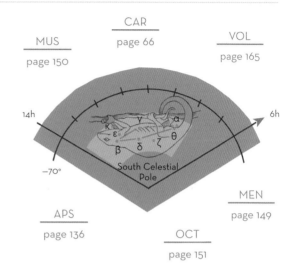

CIRCINUS

CIR – CIRCINI / THE COMPASSES

One of the smallest constellations, this is another creation of Abbé Nicolas de Lacaille during his stay in Cape Town in 1751–2. Its faint stars could have belonged more naturally with the neighbouring, better-known constellations. Lacaille named this figure in honour of the compasses used by surveyors, placing it beside another of his inventions, Norma, the surveyor's level. Circinus is most easily located as a group of minor attendants flanking Rigil Kentaurus (α Cen) to the south and east. It is visible from the Tropics and the South. α Cir (3.2, white) has a companion star of magnitude 8.6. Circinus' midnight culmination is in August.

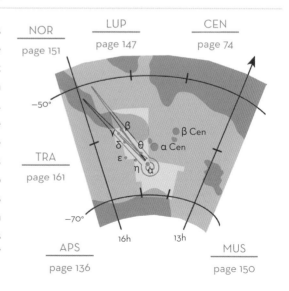

COLUMBA

COL – COLUMBAE / THE DOVE

The imagery of Columba appears to have evolved through Bayer and the Dutch cosmographer Petrus Plancius (16th century). These minor stars, north of Puppis, the stern, led to the idea of a bird flying in attendance, as in the Biblical story of the dove that flew from Noah's Ark. A parallel Greek reference is to the dove that guided the Argonauts between the Symplegades, rocks that crushed ships at the mouth of the Black Sea. Columba culminates at midnight in mid-December.

Phact ("ring dove") is the 3rd-magnitude, blue-white α star; β (yellow, 3rd magnitude) is Wasn ("weight", as in a sounding line). Both stars are bringers of good news.

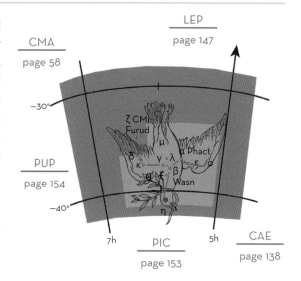

CMA
page 58

LEP
page 147

PUP
page 154

PIC
page 153

CAE
page 138

7h   5h

COMA BERENICES

COM – COMAE BERENICES / BERENICE'S HAIR

This faint group, best observed through binoculars, lies north of Virgo and east of Leo, and it once formed part of the lion's tail. Gerard Mercator defined the present-day constellation in 1551. Its midnight culmination is on April 2. Classical tradition relates these stars to the legend of Queen Berenice of Egypt, who cut off her hair as a sacrifice to the goddess Venus after her husband's safe return from battle. It is also regarded as Thisbe's veil (see page 99).

The α star is named Diadem, which refers to a wreath of jewels in Berenice's hair. Approximately 5° to the west of β Com lies the North Galactic Pole.

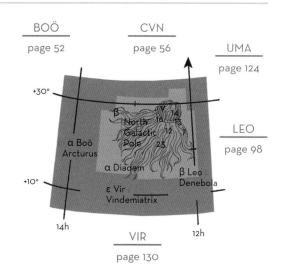

BOÖ
page 52

CVN
page 56

UMA
page 124

LEO
page 98

VIR
page 130

14h   12h

CORVUS

CRV – CORVI / THE CROW

This ancient constellation is grouped with Crater and Hydra. The crow or raven appears as a trapezium of four stars, southwest of Spica (α Vir). Its midnight culmination occurs around March 28.

Apollo sent the raven with a chalice to fetch the waters of life, but the bird loitered by a fig tree waiting for the fruit to ripen. He claimed that a watersnake had delayed him. Apollo cursed the bird with eternal thirst, fixing him in the sky with Hydra guarding the cup — thus fulfilling the lie.

Alchiba (α; 4.0; white) is from the Arabic for the group; it may once have been brighter and red. β (Kraz; yellow) and λ (Gienah; blue-white) are both magnitude 3.

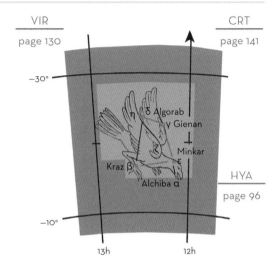

CRATER

CRT – CRATERIS / THE CUP

This ancient constellation is found directly south of Denebola (β Leo). It represents the chalice of Apollo, and is the cup carried by Corvus (see above) on his abortive mission to fetch the waters of life. Roman mythology did not limit the ownership of the cup to Apollo, and it has been variously attributed to Bacchus (in Greek myth, Dionysius), Hercules (Heracles) and the Greek Achilles, among others. Crater culminates at midnight around March 12.

The α star (magnitude 4.2; orange-yellow) is named Alkes, which means "shallow basin", or wine-vessel; it was the Arabic name for the whole constellation.

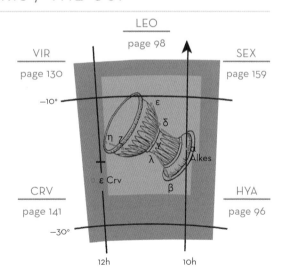

DELPHINUS

DEL – DELPHINI / THE DOLPHIN

Delphinus swims east of the bright star Altair (α Aql). Just above the equator, it is visible worldwide, other than from the Antarctic. It culminates at midnight on July 31. For the Greeks it was the "sacred fish". In India these were fortunate stars, associated with the porpoise. Early Arabs termed them the "precious stones", but later borrowed the dolphin from the Greeks. The rectangle of four main stars is called "Job's Coffin".

The names of the α and β stars, Sualocin and Rotanev respectively, emerged in the 1814 Palermo catalogue. Backward they read, "Nicolaus Venator", a Latinized version of Niccolo Cacciatore, assistant at the Palermo Observatory.

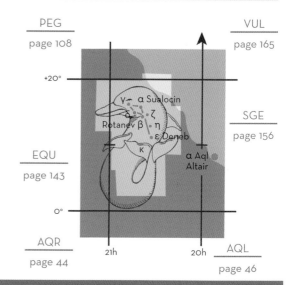

DORADO

DOR – DORADUS / THE GOLDFISH

This Southern constellation was introduced by the Dutch navigators Pieter Keyser and Frederick de Houtman in 1595–7. Occasionally it was the "swordfish". Midnight culmination is December 17.

Located southwest of the brilliant star Canopus (α Car), Dorado can be spotted by the Large Magellanic Cloud (LMC) at its southern border. Recorded by the explorer Ferdinand Magellan in 1519, this appears as a patch 6° in diameter. A mini-galaxy, it is a satellite of our own system, 169,000 light years distant. It contains around 10,000 million stars. The Tarantula Nebula, NGC 2070, lies within the LMC.

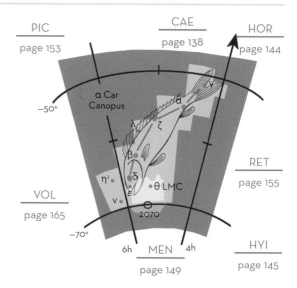

EQUULEUS

EQU – EQUULEI / THE FOAL OR LITTLE HORSE

The second smallest constellation, this seems to have been created by the Greek astronomer Ptolemy (2nd century CE). It is observable from both hemispheres, everywhere except the Antarctic. A faint trapezium lying southeast of Delphinus, between the dolphin and Enif (ε Peg), it culminates at midnight in early August.

Its α star (3.92; yellow) is known as Kitalpha, derived from the Arabic word meaning "part of the horse". It has been connected with the Gemini twins, Castor and Pollux. In one account, Equuleus was given to Castor by the god Hermes. Alternatively, it was given to Pollux (in Greek myth, Polydeuces) by Hera.

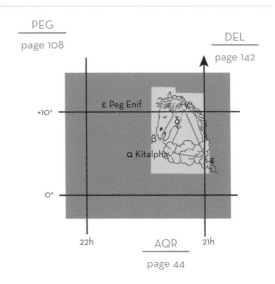

FORNAX

FOR – FORNACIS / THE FURNACE

Fornax is a Southern constellation lying in a bend on the west bank of Eridanus. A creation of Abbé Nicolas de Lacaille in 1751–2, it was originally *Fornax Chemica*, the "chemical furnace". The astronomer Johann Bode (1747–1826) later sought to dedicate it to the French chemist Antoine Lavoisier (1743–1794). It is in theory fully visible up to 50° North, but as it comprises 4th- and 5th-magnitude stars it is in practice invisible low on the horizon: it can be properly observed only at the Tropics or in the Southern hemisphere. It culminates at the beginning of November. Both the α (magnitude 3.87) and β (4.46) stars are yellow.

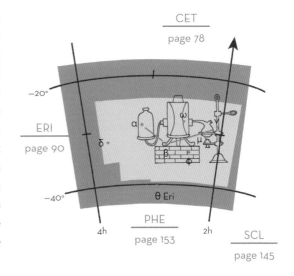

GRUS

GRU – GRUIS / THE CRANE

Grus, lying south of the bright star Fomalhaut (α PsA), becomes lost from view at middle-latitudes North, but is distinctive from the Southern hemisphere. Its midnight culmination is August 28. For the Arabs, Piscis Austrinus covered these stars. Grus was created in its present form in 1603 by the German astronomer Johann Bayer, best known for his Greek lettering of the stars. In the Middle Ages, the star-group was known as *Phoenicopterus*, the "flamingo".

α (1.74; blue-white) is named Alnair, meaning "bright one"; β (red) is a variable star (2.0–2.3); δ and μ are two pairs of unrelated stars visible as doubles to the naked eye.

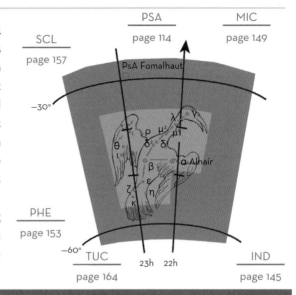

SCL — page 157
PSA — page 114
MIC — page 149
PHE — page 153
TUC — page 164
IND — page 145

HOROLOGIUM

HOR – HOROLOGII / THE CLOCK

Abbé Nicolas de Lacaille created this figure in 1751–2 from a thread of stars on the east bank of Eridanus, near Achernar (α Eri). Originally *Horologium Oscillatorium* ("the pendulum clock"), it was dedicated to the invention by Dutch scientist Christiaan Huygens in the 1650s. The Victorian scholar R.H. Allen, in his *Star Names*, cites an isolated occasion when these stars were known as Horoscopium, the "horoscope".

Horologium can be seen from the Tropics and the Southern hemisphere; it culminates at midnight during November. The α star (3.86; yellow) marks the bottom of the pendulum. β (4.99; white) is by the clock's face.

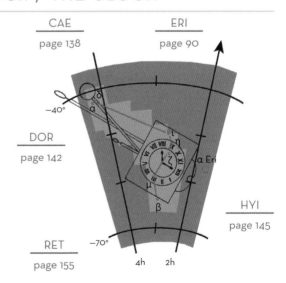

CAE — page 138
ERI — page 90
DOR — page 142
HYI — page 145
RET — page 155

HYDRUS

HYI – HYDRI / THE LESSER WATERSNAKE

Hydrus, the lesser or male watersnake, is a modern constellation of the Southern hemisphere first published by Johann Bayer in 1603. Bayer intended it as a complement to Hydra, which is an ancient constellation (see pages 145–6). The head of Hydrus touches Octans at the South Pole, while the tail reaches up close to Achernar (α Eri). It culminates at midnight in late October.

α (2.86; white) is a hydrogen star of very high surface temperature, approximately 20,000°K; β (2.80; beautiful clear yellow) is the constellation's lucida and is noteworthy by being the closest bright star (although not the closest star) to the South Celestial Pole (approximately 12° distant).

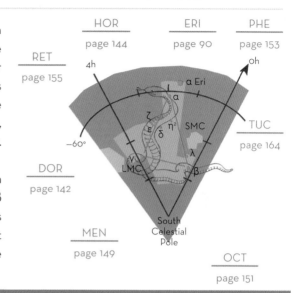

INDUS

IND – INDI / THE INDIAN

This faint Southern constellation was created by the Dutch navigators Pieter Keyser and Frederick de Houtman during 1595–7; but credit has also been given to Johann Bayer. Southwest of Grus, it stretches to Octans at the South Celestial Pole. Midnight culmination is around August 12. The figure may commemorate the American Indians, perhaps of Tierra del Fuego and Patagonia, encountered by Portuguese explorer Magellan in the early 16th century.

α (3.11) and β (3.65) are both orange stars. ε (4.69; yellow) is not unlike the Sun, but smaller and cooler; one of our closest stellar neighbours, it lies at 11.2 light years.

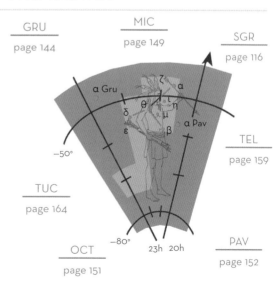

LACERTA

LAC – LACERTAE / THE LIZARD

South of Cepheus, between Andromeda and Cygnus, lie the 4th- and 5th-magnitude stars that form Lacerta. It was created by the Polish astronomer Johannes Hevelius in 1687. Hevelius alternatively offered the name Stellio, the "stellion", a newt with star-like markings found in the Mediterranean. Other short-lived names have included Sceptrum (the sceptre); and the Hand of Justice, by astronomer Augustine Royer in 1679, to honour the French king, Louis XIV. The lizard is usually depicted pointing to the north, with the α (3.77; blue-white) and β stars marking its head. The figure fades from view at middle-latitudes South. Midnight culmination is around August 28.

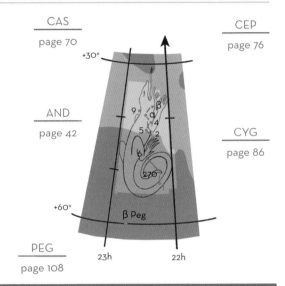

LEO MINOR

LMI – LEONIS MINORIS / THE LITTLE LION

This insignificant Northern constellation lies between Ursa Major and Leo. It was introduced in 1687 by the Polish astronomer Johannes Hevelius, but has not attracted enthusiasm from celestial cartographers. The lucida is 46 LMi (3.8; orange).

The zodiac at the temple of Dendera in Egypt (see image, page 13) shows Cancer here. In R.H. Allen's *Star Names*, these stars, with others in Ursa Major's feet (notably the pairs ν and ξ, and λ and μ UMa), constitute a distinct scarab (the Nile dung beetle; see page 54) with its front legs stretched out. Leo Minor disappears at middle-latitudes South. Midnight culmination is on February 23.

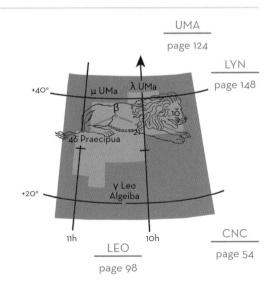

LEPUS

LEP — LEPORIS / THE HARE

Lepus has been known since ancient times. It is easily found from its location at the feet of the giant Orion, south of Saiph and Rigel (κ and β Ori). Although distinct, the hare is overshadowed by its hunter. The hunter's dog Canis Major lies immediately to the east, poised ready to leap at its prey.

The Arabs initially saw these stars as the throne of Orion, but later adopted the alternative Greek interpretation of the hare. D'Arcy Thompson, a 19th-century Scottish ornithologist, explains the location of this constellation by the legend that hares absolutely detest the sound of ravens, which is reflected in the risings and settings of stars: as Corvus the crow rises, Lepus the hare sets, scurrying into the earth to find security away from the crow.

Lepus culminates at midnight in mid-December. Lying south of the equator it is visible for all locations outside the Arctic circle. The α star (2.58; white) is Arneb, from the Arabic for "hare". Nihal (β; 2.84; yellow) is derived from the Arabic for a source of water. The stars of Lepus were sometimes seen by the Arabs as four camels, slaking their thirst at the Milky Way, and were together referred to as *Al Nihal*; β has now adopted this name. λ (3.6; yellow) is an interesting binary for observation with binoculars, with an orange companion of magnitude 6.2; the system lies 27 light years distant.

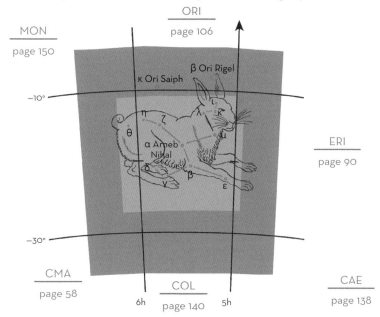

ORI
page 106

MON
page 150

ERI
page 90

CMA
page 58

COL
page 140

CAE
page 138

−10°

−30°

6h

5h

β Ori Rigel

κ Ori Saiph

α Arneb
Nihal

THE MINOR CONSTELLATIONS

LUPUS

LUP – LUPI / THE WOLF

Lupus, on the northern shore of the Milky Way, south of Libra, can be found easily by the Southern-hemisphere observer from the brilliant α and β stars in Centaurus. Its midnight culmination is in early May. α (2.30; a blue giant) lies 620 light years distant.

Classical authors described Lupus as a beast impaled on the centaur's spear, offered as a sacrifice to the gods on their altar, the nearby Ara (see page 137). Alternatively, the stars represented King Lycaon of Arcadia, who offered a dish of human flesh to Zeus (in Roman myth, Jupiter), and was turned into a wolf. The Arabs saw a lioness or a spotted leopard.

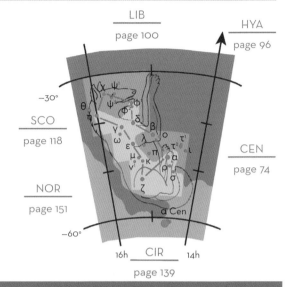

LYNX

LYN – LYNCIS / THE LYNX

Lynx stretches over a barren region of the Northern sky between Auriga and Ursa Major, north of Gemini. The figure disappears at middle-latitudes South, but even under ideal conditions in the North it is easily overlooked; it has just one star as bright as 3rd magnitude, α (3.13), a red giant 150 light years distant. The figure owes its existence to Polish astronomer Johannes Hevelius. He gave it this name in 1687 because only those with keen eyes like a lynx would find it. The figure was occasionally seen as a tiger, its faint stars as markings on its back, but this idea has not survived. Midnight culmination is January 19.

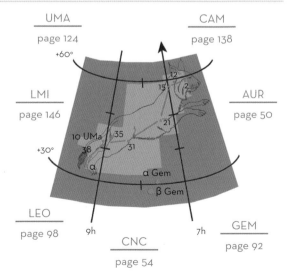

MENSA

MEN – MENSAE / THE TABLE MOUNTAIN

Mensa, meaning "table", is a Southern circumpolar constellation, and one of the 14 constellations of Abbé Nicolas de Lacaille (1751–2). As he watched the skies from Table Mountain (which dominates the Cape of Good Hope), he honoured the mountain with its own stars. The Large Magellanic Cloud is the best way to find Mensa – a portion of it overlaps from Dorado. It has been thought that Lacaille compared this with the cloud-cap of Table Mountain. Mensa should be visible from a few degrees North of the equator to the South Celestial Pole; but the faintness of its loop of four main stars, α, β, λ and η, renders it a challenge for the naked eye.

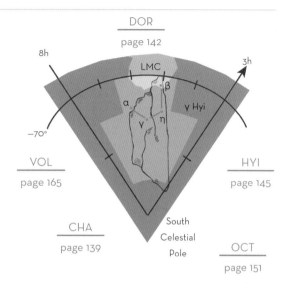

MICROSCOPIUM

MIC – MICROSCOPII / THE MICROSCOPE

Microscopium is a small Southern constellation to the south of Capricornus. A number of the celestial figures of the 17th and 18th centuries celebrate the progress of science. The indefatigable Abbé Nicolas de Lacaille, creator of Microscopium, is foremost in this spirit of the "age of Enlightenment", having created 14 of the modern constellations. It is named in honour of a revolutionary invention, decisive in the advance of science, yet it is made up of faint 5th-magnitude stars, which make it very difficult to observe. The lucida is the λ star, a dim magnitude 4.7, and yellow. Its midnight culmination is around August 4.

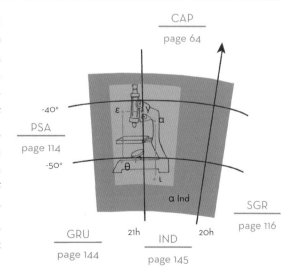

MONOCEROS

MON – MONOCEROTIS / THE UNICORN

Named by the Dutch astronomer Petrus Plancius in 1613, Monoceros straddles the Milky Way at the equator, within the triangle of Sirius (α CMa), Procyon (α CMi) and Betelgeuse (α Ori). Its magical horn lies close by Orion's eastern shoulder. The 13th-century *Bestiaire Divin de Guillaume* tells us that when a virgin is placed in the unicorn's haunts, the beast will lie still and can be caught by a hunter. The unicorn is said to represent Christ, its horn the Gospel of Truth.

In this star-field, the Rosette Nebula (NGC 2237 and 2244) is just visible to the naked eye. NGC 2264, the Cone Nebula, includes the 5th-magnitude star σ Mon.

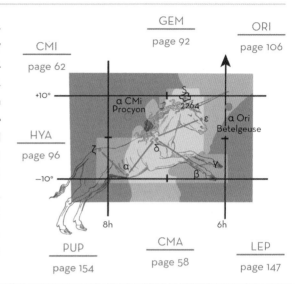

GEM
page 92

ORI
page 106

CMI
page 62

HYA
page 96

PUP
page 154

CMA
page 58

LEP
page 147

MUSCA

MUS – MUSCAE / THE FLY

A small but distinct Southern-hemisphere constellation, Musca is easily found from its location on the Milky Way immediately south of Crux. It culminates at midnight at the end of March. It was invented by the Dutch navigators Keyser and de Houtman in the 1590s, originally under the name of Apis, the bee. As this could easily be confused with Apus (see page 136), it later became known as Musca Australis, the Southern Fly, to distinguish it from the now defunct Musca Borealis, the Northern Fly, comprising a cluster of stars in Aries. α Mus (magnitude 2.7) is blue-white. β (3.05) is a binary pair (3.7 and 4.0).

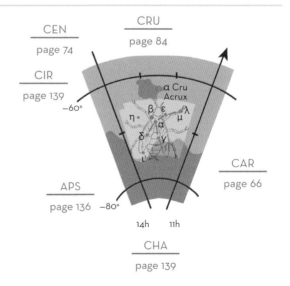

CRU
page 84

CEN
page 74

CIR
page 139

APS
page 136

CAR
page 66

CHA
page 139

NORMA

NOR – NORMAE / THE LEVEL

This Southern constellation, lying on the Milky Way, was created by Abbé Nicolas de Lacaille in 1751–2 from undesignated stars in Lupus (to the northwest), Ara (east) and Scorpius (north). It has only one star brighter than 5th magnitude. He named it to honour the set square used by ships' carpenters and navigators on their voyages of discovery. The pair of stars, including α, originally forming the northern end of the level now lie in Scorpius, a few degrees to the west of ε Sco. γ¹ and γ² (5.0 and 4.0, yellow) make a fascinating pair; γ² lies at 145 light years, but γ¹, a yellow supergiant, is 10,000 light years distant. Midnight culmination occurs around May 19.

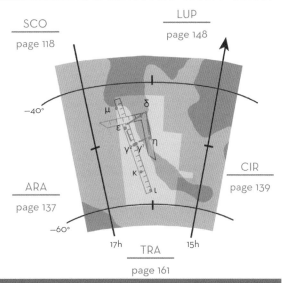

OCTANS

OCT – OCTANTIS / THE OCTANT

Octans, another creation of Abbé Nicolas de Lacaille, in 1751–2, was named in honour of the octant, a forerunner of the sextant, invented in 1730 by John Hadley and used to determine stellar positions. Octans is circumpolar, and the South Celestial Pole lies within its bounds (see page 155).

σ Oct (magnitude 5.4; white) is the Pole Star of the South, lying approximately 1° from the pole itself. It has little navigational use as it is at the extreme edge of naked-eye observation, and requires ideal viewing conditions. The Northern hemisphere Pole Star, Polaris, by comparison, is more than 20 times brighter.

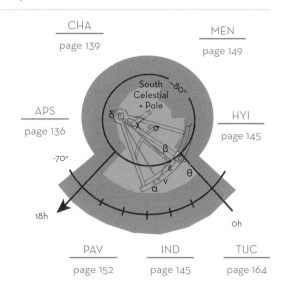

PAV – PAVONIS / THE PEACOCK

Moving south from the distinctive constellations Sagittarius and Corona Australis, we cross dim Telescopium to reach Pavo, the peacock, created by the Dutch explorers Pieter Keyser and Frederick de Houtman in 1595–7. Pavo is visible from the Tropics and the Southern hemisphere; it culminates at midnight in mid-July. The α star is called "Peacock" (1.9; blue-white).

There is a curious thread of association in Greek myth that may have appealed to the European navigators in the Southern seas; for Argus, as well as being a mythical figure linked with the peacock, is also the name of the shipwright who built the *Argo* in which Jason sailed to retrieve the Golden Fleece (see pages 66–8). The primary story, however, concerns, Zeus (in Roman myth, Jupiter) who tried to disguise an affair with the girl Io by turning her into a white heifer. His wife Hera asked for the heifer as a gift, which he could not refuse without raising suspicion. Hera placed Io, now a cow, under the care of the watchman Argus, called Panoptes ("all-seeing"), who had one hundred eyes. To save Io, Zeus sought the help of wily Hermes, who took his lyre, played Argus music and told him stories. Once the last of the watchman's one-hundred eyes had fluttered into sleep, Hermes slew Argus, cutting off his head. Hera set his eyes into the tail of the peacock.

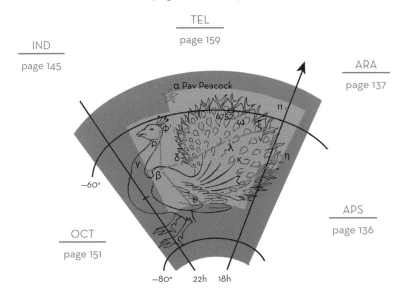

TEL
page 159

IND
page 145

ARA
page 137

OCT
page 151

APS
page 136

PHOENIX

PHE – PHOENICIS / THE PHOENIX

This figure is one of a group of four constellations named after birds in this region of sky (Pavo, Tucana and Grus are the others). It owes its genesis in 1595–7 to the Dutch navigators Pieter Keyser and Frederick de Houtman. Lying east of Eridanus, it is found from nearby Achernar (α Eri). Visible from the Tropics and the Southern hemisphere, it culminates at midnight on October 4. The name Ankaa for α (2.39; yellow) is of uncertain origin.

The Phoenix, a mythical bird that periodically consumed itself in fire but then rose anew from the ashes, was associated with immortality and the secrets of alchemy. For the early Arabs the stars formed a boat.

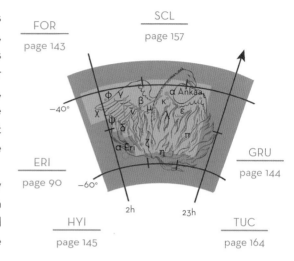

PICTOR

PIC – PICTORIS / THE PAINTER'S EASEL

A Southern constellation created by the Abbé Nicolas de Lacaille during his stellar observations of the South in 1751–2. Faint and insignificant, it covers some minor stars lying between the star Canopus (α Car), the constellation Dorado and the Large Magellanic Cloud to the south. It culminates at midnight around December 16 and can be observed from the Tropics and the Southern hemisphere. α is a white star of magnitude 3.27. Remarkably, β (3.85; blue-white), at 59 light years, is thought to be a planetary system in the making – a photograph taken in 1984 revealed that it is surrounded by a disk of dust and gas.

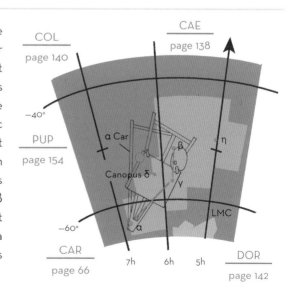

THE MINOR CONSTELLATIONS

PUP — PUPPIS / THE STERN

Puppis is the stern or "poop" of the ship Argo Navis, an enormous Southern constellation (see pages 66–8), divided into three manageable sections: Puppis, Carina (the keel), and Vela (the sails). The largest of the three, Puppis lies partly on the Milky Way in a region dense with stars and ideal for a sweep with binoculars. It can be located from Canopus in Carina to the southwest, and Sirius in Canis Major to the northwest.

The stern is the most northerly part of the ship, and the only part that can be seen from the Northern hemisphere. From middle-latitudes North it is possible to see a group of stars around Azmidiske and Markeb, representing a flag flying above the ship's poop. Below 39° North it is possible to see the whole figure, although it is faint when low on the horizon. Puppis culminates at midnight in early January.

As with the other portions of Argo Navis, the Bayer-lettering follows the original constellation, so Puppis has no star designated α, β, λ or δ. Naos (ζ; 2.25; brilliant blue-white), derived from the Arabic for "ship", is a supergiant, 1,500 light years distant and one of the hottest stars known (surface temperature approx. 63,000°F/ 35,000°C). Azmidiske (ξ; 3.34; yellow) lies at 650 light years; its unrelated companion (5.3; orange) lies at 280 light years. This pair can be easily resolved with binoculars.

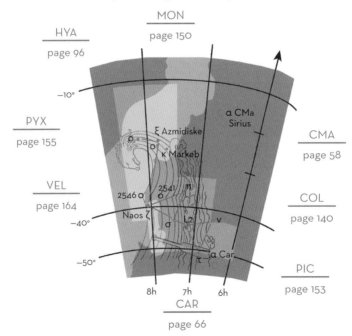

PYXIS

PYX – PYXIDIS / THE COMPASS

To the east of Puppis lies Pyxis, the compass. Although partly on the Milky Way, it contains few objects of interest. Its creator, Abbé Nicolas de Lacaille, who in 1751–2 mapped the stars of the Southern hemisphere, was honouring the invention of the magnetic compass used by seafarers, which makes some sense of its placing close to Argo Navis (see pages 66–8). The original title was *Pyxis Nautica*, the "nautical box". Although they are obscure, these stars were known in the 2nd century CE to Ptolemy, who created in the same area Malus, the mast, as a subdivision of Argo Navis. The α star is magnitude 3.68 and blue-white. Pyxis culminates at midnight in early February.

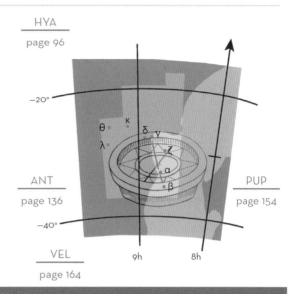

HYA
page 96

ANT
page 136

PUP
page 154

VEL
page 164

RETICULUM

RET – RETICULI / THE NET

This small Southern star-group, culminating at midnight on November 19, is halfway between Achernar (α Eri) and Canopus (α Car). It was published posthumously for de Lacaille in 1763 as *Reticulum Rhomboidalis*, "the rhomboid net". He had in mind a reticle or grid used in a telescope eyepiece to provide scale and location. R.H. Allen, in *Star Names*, gives the real credit for the figure to Isaak Habrecht of Strasbourg, who published it under the name Rhombus before de Lacaille. α is magnitude 3.35 and yellow; β (3.85) is orange. ζ (5.24) is a wide double of almost identical yellow stars, ζ² (4.98) and ζ¹ (5.5); both are like our Sun.

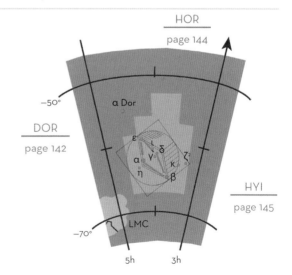

HOR
page 144

DOR
page 142

HYI
page 145

SGE — SAGITTAE / THE ARROW

For the Greeks this constellation was even smaller than our figure, covering just 4° of arc. Despite its smallness this is a distinctive grouping, consistently described since ancient times. Easily located, lying across the Milky Way around 10° north of the brilliant star Altair (α Aql), Sagitta is visible worldwide except from the Antarctic. It culminates at midnight on July 16.

Sagitta has been described as an arrow by the Hebrews and the Persians, as well as by the authors of Greek and Roman antiquity. It is shown speeding west to east, or vice versa, often depending upon the mythological slant being given. It has represented arrows of various myths, but is regularly linked with the

story of Heracles (in Roman myth, Hercules), whose constellation lies directly west of Sagitta. In one account the arrow is that used by Heracles to save Prometheus from having his liver pecked out by Zeus' eagle (see pages 46–7). In another Sagitta is the arrow with which the hero, as one of his Labours, killed the birds that plagued the people of Stymphalus — Cygnus, Aquila and (as the vulture) Lyra.

The stars α and β (both 4.37; yellow giants) form an interesting pair, closely aligned like the flights of an arrow. They are similar distances from Earth, 610 and 640 light years respectively. The lucida is λ (3.47; orange), 175 light years distant.

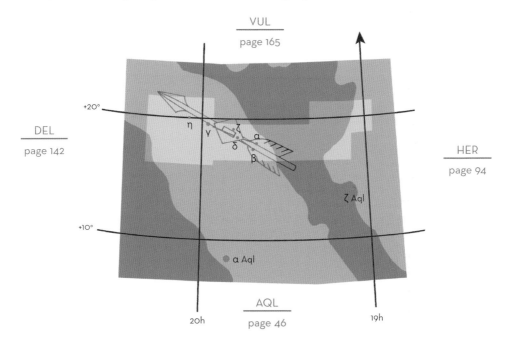

SCULPTOR

SCL – SCULPTORIS / THE SCULPTOR

Created by the Abbé Nicolas de Lacaille in 1751–2, Sculptor is arguably a more useful creation than some of his 13 other efforts, as it is in an area of previously-unassigned faint stars, east of Piscis Austrinus. A few degrees northeast of the α star (4.31; blue-white) is the South Galactic Pole. The Abbé originally named this constellation *L'atelier du sculpteur*, the "sculptor's workshop"; Latin variations on this name occur in some 19th-century star maps. Sculptor is best viewed from the Tropics or the Southern hemisphere, although it can be seen under good conditions up to approximately 50° North. Midnight culmination is around September 26.

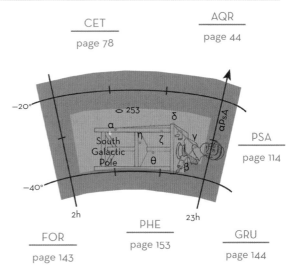

SCUTUM

SCT – SCUTI / THE SHIELD

Scutum lies on the Milky Way between Aquila and Serpens Cauda. Created by Johannes Hevelius (1611–1687) in honour of King John Sobieski III of Poland, it was originally called "Sobieski's Shield", pictured as the king's coat of arms with a cross. Scutum is visible everywhere outside the Polar circles. Midnight culmination is July 1.

M11, the Wild Duck Cluster, has roughly 200 stars, 5,600 light years distant. Just visible to the eye, it is easily seen with binoculars as a hazy patch, half as wide as the Full Moon. In 1844 astronomer William Smyth described it as "a flight of wild ducks in shape"; and the name has stuck ever since.

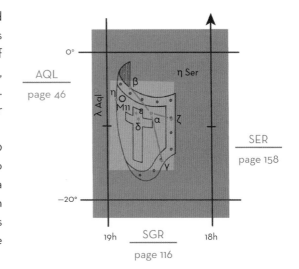

SERPENS

SER – SERPENTIS / THE SERPENT

The Serpent coils around Ophiuchus, the serpent holder, who cuts the creature into two apparently unconnected parts, making Serpens unique among the constellations. Caput and Cauda are the Head and the Tail of the Serpent respectively. The tail lies on the Milky Way at the equator, west and a little south of Altair (α Aql). The neck and head of the serpent rear up, west of Ophiuchus, toward Corona Borealis. The head is marked by a compact triangle of stars (β, λ and κ). Part of the serpent can be seen anywhere on Earth. However, none of its stars is brighter than 3rd magnitude, so that at high Southern latitudes, northerly reaches of the serpent, at the head, require ideal viewing conditions. Midnight culmination occurs in the third week of May for the head, a month later for the tail.

Serpens is best known in relation to the healer Aesculapius (see pages 104–5). The power of medicine is said to be that of the serpent's venom, which can kill or cure, depending upon how it is used. The shedding of the snake's skin was from early times representative of rejuvenation.

Unukalhai (α; 2.65; orange) means "serpent's neck". θ (4.06) is a beautiful binary system, at 105 light years, comprising twin white stars (4.6 and 5.0), separable with good binoculars; its name, Alya, may derive from the Arabic *Al Hayyah*, "the snake".

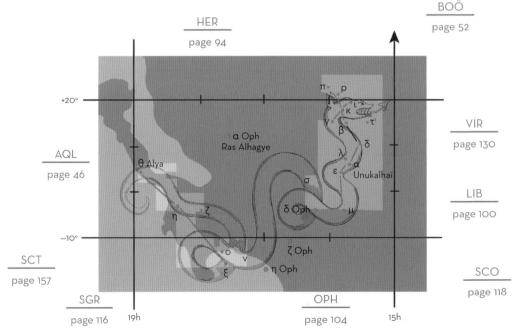

HER
page 94

BOÖ
page 52

AQL
page 46

VIR
page 130

LIB
page 100

SCT
page 157

SCO
page 118

SGR
page 116

OPH
page 104

+20°

α Oph
Ras Alhagye

θ Alya

Unukalhai

δ Oph

ζ Oph

η Oph

−10°

19h

15h

SEXTANS

SEX – SEXTANTIS / THE SEXTANT

The development of astronomy during the 17th century meant that unconsidered areas of sky began to receive attention. The faint Sextans (originally *Sextans Uraniae*, the "sextant of Urania") was created from border areas between Hydra and Leo by Johannes Hevelius (1611–1687). Urania was the muse of astronomy. α Sex (4.49; blue-white) lies almost exactly on the equator, 12° south of Regulus (α Leo). Sextans culminates at midnight around February 22. The constellation's name commemorates the instrument, used to measure stellar positions. Hevelius had in mind his own sextant, which had been destroyed in a fire at his observatory in September 1679.

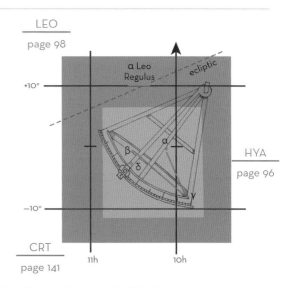

TELESCOPIUM

TEL – TELESCOPII / THE TELESCOPE

Devised by de Lacaille during his observations in 1751–2, Telescopium lies south of Sagittarius and Corona Australis. It culminates at mdinight around July 10. Close by its southeast corner is Peacock, Pavo's lucida. α is magnitude 3.51 and blue-white.

It is no surprise to find represented in the heavens this revolutionary astronomical invention. Observation tubes had been used since ancient times; but the first clear evidence of a telescope, with magnifying lenses, appears in a letter dated 1608 from a Committee of Councillors in Zeeland in the Netherlands. By 1610 Galileo had achieved magnifications of 20–30 times.

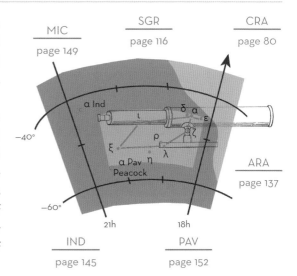

THE MINOR CONSTELLATIONS

TRI – TRIANGULI / THE TRIANGLE

Triangulum lies 10° north of the head of Aries. The name Metallah (α; 3.41; white) is from the Arabic for "triangle". β (3.00) is white. Midnight culmination is October 23. To the Greeks this was the letter δ (delta), from which it came to represent a river delta, especially the Nile's, hence its naming as "the river's gift". It also represents Sicily (Trinacia to the Greeks) from the island's triangular shape; Sicily was sacred to Demeter, and it was from here that Persephone was taken by Hades (see page 131). M33, a spiral galaxy at around 2.7 million light years, is one of the most remote objects visible to the naked eye. In ideal conditions it is a patch larger than the Moon.

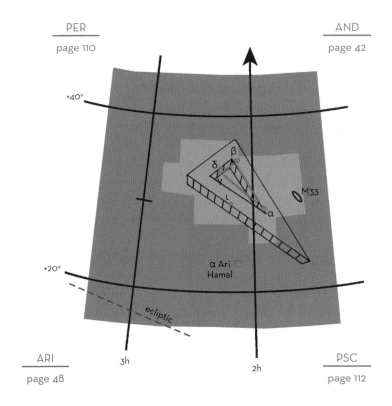

PER
page 110

AND
page 42

ARI
page 48

PSC
page 112

TRIANGULUM AUSTRALE

TRA – TRIANGULI AUSTRALIS / THE SOUTHERN TRIANGLE

Triangulum Australe is one of the keys to orientation in the Southern skies. It is located on the southern side of the Milky Way, east and south of Rigil Kentaurus (α Cen). The name Atria (1.92; orange) is occasionally used for α TrA, but its origin is obscure. The triangle is formed by this star with β (2.85; white) and λ (2.89; blue-white).

The first reference to the figure appears to be that of the Italian navigator Amerigo Vespucci in 1503, although it did not appear in a star atlas for a further one hunded years. The constellation fades entirely from view north of the Tropics. It culminates at midnight around May 23.

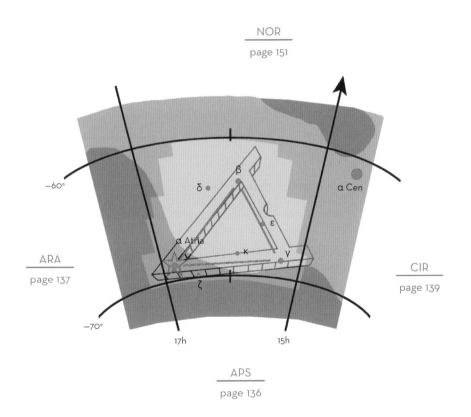

NOR
page 151

ARA
page 137

CIR
page 139

APS
page 136

THE SOUTH POLE

SIGNPOST MAP 3

The map shows the Southern hemisphere sky immediately around the South Celestial Pole at midnight local mean time (1am with Daylight Saving) on December 21, or 10pm on January 21 (11pm with Daylight Saving). The nearest star to the South Pole is σ Oct, but it is a dim magnitude 5.4, and is unhelpful for locating the pole. The α and β stars of Centaurus, Rigil Kentaurus and Hadar, the "pointers", form an approximate line to the transverse beam of the Southern Cross, Crux. The vertical axis of the Cross covers an arc of 6° from λ Cru through α Cru. This line extended points fairly closely to the pole. Centaurus is another good pointer for nearby star-groups: a short hop over dim Circinus beneath the centaur's feet will take us to Triangulum Australe — one of the brightest among the Southern circumpolar constellations.

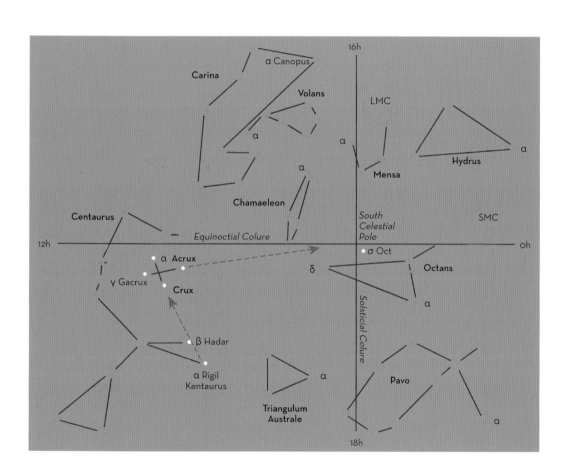

TUCANA

TUC – TUCANAE / THE TOUCAN

This Southern circumpolar constellation was introduced by Dutch navigators Pieter Keyser and Frederick de Houtman, during 1595–7. It is readily found from Achernar (α Eri), northeast of Tucana. In the earliest surviving illustration (by Johann Bayer, 1572–1625), the bird carries in its large beak the stem of a plant, marked by α (2.86; orange; now the beak itself) and δ, and is perched on the Small Magellanic Cloud (SMC). This is a satellite of the Milky Way (at 190,000 light years). To the naked eye it is a tadpole-shaped patch, 3.5° across; binoculars reveal a central blaze. 47 Tuc (NGC 104) is a globular cluster, appearing as a hazy 4th-magnitude star.

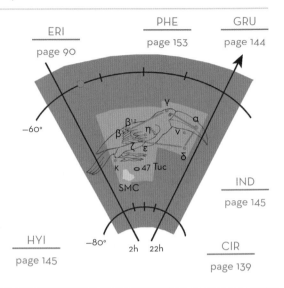

ERI
page 90

PHE
page 153

GRU
page 144

IND
page 145

HYI
page 145

CIR
page 139

VELA

VEL – VELORUM / THE SAILS

Vela once formed the sail of the ship Argo Navis (see pages 66–8), which de Lacaille first separated in 1763. Vela lacks α and β stars, which remain in Carina, so its lucida is δ (1.96; white). λ (2.21; orange-red) is named Alsuhail, from the Arabic for the "bright one of the weight". κ and δ Vel, with ι and ε Car, form the "False Cross", so called as it can be mistaken for Crux (see pages 84–5); in fact, it is a useful locator for Vela and Carina. The stellar cluster IC 2391 is visible with the naked eye. It comprises 50 stars at 590 light years, encircling o (blue-white). Vela disappears from 33° North upward. It culminates at midnight around February 13.

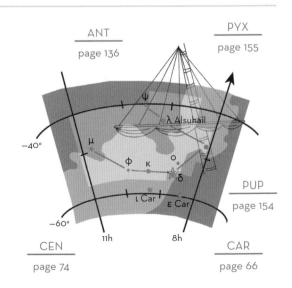

ANT
page 136

PYX
page 155

PUP
page 154

CEN
page 74

CAR
page 66

VOLANS

VOL – VOLANTIS / THE FLYING FISH

This Southern constellation yields a few 4th- and 5th-magnitude stars. It was created by the Dutch navigators Pieter Keyser and Frederick de Houtman in 1595–7. Sailors were impressed by the fish with large pectoral fins, which enabled these creatures to skim across the surface of the ocean for hundreds of yards. The constellation's full name was Piscis Volans; the modern name means simply "flying". To the east and north, Volans borders Carina; it lies south of Avior (ε Car), west of Miaplacidus (β Car). Volans is invisible north of the Tropics. Its lucida is β (3.77; orange), marginally brighter than α (4.00; white). Volans culminates at midnight in mid-January.

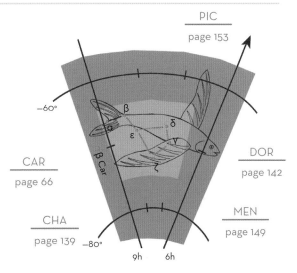

PIC
page 153

CAR
page 66

DOR
page 142

CHA
page 139

MEN
page 149

VULPECULA

VUL – VULPECULAE / THE FOX

In 1687 the Polish astronomer Johannes Hevelius created *Vulpecula cum Ansere*, the "little fox with the goose", on the Milky Way between Sagitta and Cygnus. Its brightest star, α, is a mere 4.4; it is a red giant (at approximately 250 light years). Binoculars reveal α nearby, unrelated orange giant: 8 Vul (5.8). Through binoculars or a small telescope, M27, the Dumbbell Nebula, can be seen in Vulpecula as a dumbbell-shaped, green glowing mist. Although theoretically visible everywhere north of 60° South, Vulpecula requires good viewing conditions, clear of the horizon. It culminates at midnight around July 25.

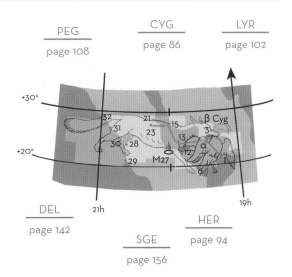

PEG
page 108

CYG
page 86

LYR
page 102

DEL
page 142

HER
page 94

SGE
page 156

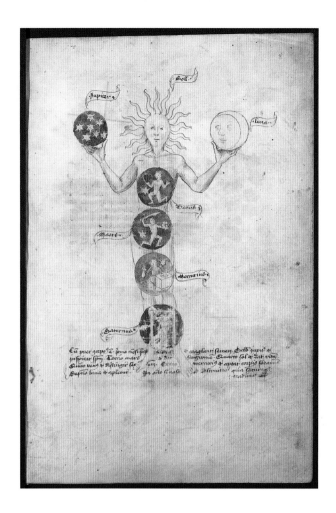

A medieval illustration in which parts of the body represent different planets. The head is the Sun. In the left hand is the Moon, and in the right Jupiter. Venus, at the chest, holds a heart as the planet of love. Beneath is Mars, the planet of war, brandishing a shield and a sword. Then Mercury, sitting at a desk holding a bag of coins. Finally comes Saturn, who represents age and time.

The Titan Atlas holding aloft a representation of the Earth, from a 19th-century fairground amusement. Atlas was commanded by his cousin Zeus (in Roman myth, Jupiter) to carry the weight of the world (or in later accounts the heavens) on his shoulders. In Greek mythology Earth was represented by the goddess Gaea, wife of Uranus. Their children were the Titans.

THE
WANDERING STARS

> " I saw Eternity the other night / Like a great Ring of pure and endless light, / All calm as it was bright, / And round beneath it, Time in hours, days, years / Driven by the spheres. "

Henry Vaughn (1622—1695), "Eternity" from Nature, Man, Eternity

For the ancients the "wandering stars" (in Greek, *planetes*) were the planets Mercury, Venus, Mars, Jupiter and Saturn, as well as the two "great lights" of the Sun and Moon. Modern discovery of the new planets — Uranus, Neptune and Pluto — has required us to call upon our mythic imaginations once more and name them accordingly. All these bodies amble across the fixed star-fields over the course of days and months, playing out their mythological associations in their physical properties and positions in the sky.

Although we may be sure that myths of the Sun are coeval with the human imagination, it seems that such concepts developed an enhanced status in settled agrarian cultures. Ancient Egypt at the dawn of civilization is a prime example. One of the most revered Egyptian deities was Horus, the falcon-headed child of Isis and Osiris. His symbolism may reflect a prehistoric association of the Sun as the celestial hunter, all-seeing like the sharp-eyed falcon. He was also interpreted as the great sky-god: the Sun and the Moon are his eyes.

Light and seeing seem universally to develop into the metaphor of revelation, prophecy and truth, as exemplified by the Greek Sun-god Apollo, whose oracle was at Delphi. His association with the crow or raven reflects the symbolism of the sharp-sighted bird (see page 141).

Beyond the simple observation of the Sun as light there emerges the idea of regularity, order and return, often figured by wheels and chariots, as in the story of the Sun-god Helios and his boisterous son Phaethon (see page 90).

The cyclic order of the Sun involves two symbolically parallel rhythms: the diurnal and the seasonal. The diurnal cycle is especially apparent in Egyptian myth. The Sun took on different meanings and names depending on its position in the sky. The solar disk was honoured as Aten; at rising, the Sun was the god Khephri, the scarab beetle; as it rose into the sky toward the meridian, it became the great creator-god Ra; at setting and during the night, it was Atum.

Ra was also shown in the complete cycle of the Sun, crossing the heavens in his Sun-boat. A consistent theme for the Sun-god is the key idea of rebirth – after darkness the Sun is born anew each morning, and rises to become king as it ascends to its full grandeur at midday.

The seasonal cycle is also associated with rebirth. The naturalistic interpretation of myth by the scholar James Frazer (1854–1944) convincingly argues that many myths rationalize the order of the seasons and their agricultural cycles. One of the best-known solar symbols is the sigil for Cancer (see page 18), which is directly derived from the Egyptian hieroglyph for the scarab Khephri, in seasonal rather than diurnal guise: the appearance of the scarab beetle after the Nile floods

An ivory carving showing the birth of the Egyptian Sun-god Horus. In some accounts his left eye represented the Sun, his right the Moon.

each year affirms the fertility of the land and the cycle of creation (see also page 54). This role is played out in the Tropical zodiac, which marks a turning point of the year at Northern midsummer (Southern midwinter), June 22, when the Sun enters the sign of Cancer.

Associated with the cyclic conception of the Sun is the "solar hero", a symbolic personage who undertakes a sequence of trials that represent the stages of the Sun's cycle. Classical mythographers were fond of turning these into 12 adventures, analogues to the 12 signs of the zodiac, although in practice it is difficult to make a fully consistent argument of this sort. The best known of these heroes is the Greek Heracles (see pages 94–5), with his 12 Labours.

With the development of astronomical observation, the Sun gained status as a regulator of the heavenly motions, especially those of the other planets. A passage in Shakespeare's *Troilus and Cressida* reflects the medieval conception of "the glorious planet Sol" enthroned at the heart of the grand design of heavenly "degree, priority and place, / Insisture, course, proportion, season, form". The power of rule and order is also deemed to be the power of the monarch, and it is no surprise that kings of all ages have sought to identify themselves with solar images. From the early Dynastic period the Egyptian pharaohs were identified as the "sons of Ra". But then in the 14th century BCE the pharaoh Akhenaten ("glory of Aten", the solar disk) banished all other gods and declared himself intermediary between humanity and the Sun as sole divine creator.

A millennium and a half later we find a similar development in the late Roman Empire, conceived in Mithraism, which came to Rome from Persia. The bull-god Mithras, also named Helios ("Sun"), is associated with the constellation of Taurus, occupied by the Sun each spring at this epoch (see pages 120–1). Mithraism's virtues of strength and purity made it popular with the military; it indirectly but substantially set the scene for the state cult of the Sun with its apex as the emperor, formally instituted in Rome in the 3rd century CE.

A jade bead carved as a Mayan effigy of a Sun-god. Mayan solar myth tells how the Sun tried to impress a weaving girl by carrying a deer by her house each day. The girl's grandfather forbade the union, so the Sun turned himself into a hummingbird that was shot by the grandfather. The girl nursed the Sun to health and became the Moon. In one tradition the gods send the Sun and Moon into the sky as punishment for the Moon's carnality.

The Moon's changing angles in relation to the Earth and Sun — its *phases* — are universally familiar. Commonly, we give the term New Moon to the faint crescent on the Western horizon one to two days after the astronomical New Moon (when it is dark). The auspicious occasion of the New Moon marks the first day of the month in lunar calendars.

In fact, the earliest calendars were lunar, not solar. The Moon's phases were not only the first count of time beyond day and night; they were also crucial to hunters for whom the Full Moon's light was as good as the light of day. It was in the change to settled

The crescent Moon was a symbol of good fortune. In this Inca pottery jar it hangs around the neck of a figure holding other jars.

agricultural societies that the seasons of the solar year became more significant. A clue to this prehistory lies in the role of the Egyptian Moon-god Thoth, sometimes depicted as a baboon with a crescent on his head. As Ra, the Sun-god journeyed through the night world and is said to have placed Thoth in the upper world. Thoth was put in charge of the calendar and had the difficult task of collating a lunar year of 13 months of approximately 29½ days each with the 365-day cycle of the Sun. The problem was solved using *intercalation*, the periodic insertion of a 13th lunar month.

On the whole the Moon has been regarded as female (Thoth is one exception). Its influence over the menstrual cycle (from Greek *menses*, "moon") has linked it to fertility and birth.

A frequent lunar motif is that of the Triple Goddess, in various roles, as three Fates or Norns, or three witches. The three represent the crescent phase, the Full phase, and the mysterious dark of the Moon. In Greek myth, the virgin Artemis, with her hunting bow, is like the crescent Moon; Selene is the Full Moon; and mournful Hecate is the dark Moon.

The story of Selene and Endymion evokes the mystery of the Moon's night world. Endymion was a handsome young shepherd. Desiring him, the Moon-goddess stole down from the heavens to the glade where he slept. He never awoke to see her silvery form, for he lies entranced for ever, living and warm, and the Moon comes to lie with him every night.

The Greeks knew this planet as the "star of Hermes", who was the messenger of the gods, equivalent to the Roman god Mercury. The planet stays close to the Sun, and its motion is relatively rapid. Because of its proximity to the Sun's overwhelming light, it is difficult to spot. These physical characteristics all seem to fit the role of the fleet-footed messenger. The Greek Hermes was a dextrous god of thieves and pickpockets. He was also marvellously gifted: before noon on the day of his birth he invented the magical lyre (see Lyra, page 102).

The Babylonian forerunner of Hermes was Nabu, god of wisdom. Nabu and his wife Tashmetum were the inventors of writing, and each year when the destinies of all beings were determined, Nabu engraved the decisions of the gods on sacred tablets.

Hermes was also identified with the Egyptian god Thoth, who had started life as a Moon-god (see opposite page). Thoth was a spokesman for the gods, and keeper of their records, in the same way that Hermes attended the Olympian gods. He taught mortals the arts and sciences, including hieroglyphics, with which to record his inventions. He was the first magician, and his magic formulae were said to have commanded the forces of nature. This power earned him the later title of Hermes Trismegistos, meaning "thrice-greatest Hermes". He became identified as the mythic father of the lineage of the Judaic, Christian and pagan mysteries of the European and Middle-Eastern cultures. He was known

as Mercurius in the Latin Middle Ages — part Christ-figure, part trickster, and the subtle guide of the alchemists and magicians.

In several European languages we find the Latin root of Mercurius in words such as com*merce* and *mer*chant. The Greek "Hermes" literally translated means "stone heap", perhaps relating to the custom of marking tracks and paths with occasional stones, augmented by travellers. The two classical cultures come together when we consider that, historically, travellers were often merchants, who journeyed to exchange their wares.

An engraving of Mercury, shown carrying the caduceus (a magic staff entwined by two snakes) and wearing a winged cap and winged sandals.

MERCVRII · STAT · I · AED · SABELLIS ·

For cultures that have inherited elements of Greco-Roman astrology and mythology, Mars (in Greek myth, Ares) and Venus (Aphrodite) represent the masculine and feminine, the lover and beloved. However, particularly with Venus, we should be aware of the complexities underlying these associations.

Venus is a brilliant planet, the brightest star-like object in the sky after the Sun and Moon, attaining at maximum an apparent magnitude of −4.4. Hence, many cultures select Venus for a special role. In Meso-american tradition, at its heliacal rising (its first observed rising just before the Sun), Venus

A Mayan codex showing the plumed serpent-god Quetzalcoatl, who was represented by Venus as the morning star.

was the plumed serpent-god Quetzalcoatl, its dazzling rays the spears that the god hurls at his enemies.

Mesopotamian mythology saw Venus as the goddess Ishtar, who was male as the morning star (rising ahead of the Sun) and female as the evening star (setting after the Sun). This feminine side became absorbed into the later classical interpretation of Venus as the goddess of love. To honour Ishtar's powers, both sacred and commercial prostitutes were enrolled at her temple.

Among the Greek deities, Aphrodite — like her planet, with its brilliant blue-white light — is one of the most beautiful. Her name means "born of the foam". When Cronus castrated his father Uranus and flung the genitals into the sea, Aphrodite emerged fully formed from the foam made by the semen. She was borne by the West Wind to Cyprus. The gods, struck with admiration for her beauty, then welcomed her onto their home, Mount Olympus.

Later, at a wedding celebration, the goddess Eris ("discord") tossed a golden apple inscribed "to the fairest" into the hall. Hera, Athene and Aphrodite all claimed it. To settle the dispute the Trojan prince Paris was chosen to act as arbiter. Each of the goddesses appeared to him. Hera and Athene made promises of lands and victories. Aphrodite simply promised that he should have the most beautiful of mortal women (Helen of Troy): Paris succumbed, proclaiming Aphrodite the fairest.

The red planet Mars (the "star of Ares" to the Greeks) has been consistently associated with strife and bloodshed. The Roman god Mars had high status among mercenaries and soldiers, and it is no surprise that he was highly honoured by this militaristic culture. The planet's colour is identified with fire, blood and danger.

The Greek Ares is prefigured by the Mesopotamian Nergal, a god who kills through war and fever. His descent into the underworld mirrors the story of the Babylonian goddess Ishtar's adventure into the infernal regions; and because Ishtar is identified with Venus, we find here an early suggestion of the mythological pairing of these planets. The surviving text is fragmentary, but it seems that Nergal insulted Namtar, and it was decreed that he should appear before the goddess Ereshkigal, queen of the underworld. Before he set off on this journey, the god of wisdom Ea gave him a special chair with which to ward off spells, and advised him to accept nothing from the goddess. When Nergal arrived before the queen, he refused all her offers of food and comfort. However, Ereshkigal then went to bathe and returned in a dress that allowed Nergal to glimpse her body. At first he resisted temptation, but when the performance was repeated, "he gave in to his heart's desire to do what men and women do". They spent six days together, and only on the seventh did Nergal ascend to the upper world. But Ereshkigal, in an exact parallel with the myth of Ishtar, threatened to raise the dead if her lover did not return. In a storm of passion, Nergal burst through the gates of the underworld to claim Ereshkigal as his wife.

In the words of Zeus in Homer's *Iliad* (8th or 9th century BCE), Ares is "a furious god, by nature fickle and wicked". He was despised by most gods except Eris and Aphrodite (Venus). The latter fell for his brutal passion. She had married the lame smith Hephaestus, but soon was attracted by Ares. Helios, the all-seeing Sun, reported this to Hephaestus, who forged an invisible net, in which he caught the adulterous couple. He summoned the gods, who roared with laughter at the sight.

This Arabic illustration from an 18th-century manuscript shows the god Mars on the left. He was the ruler of the zodiac sign Aries.

The noble giant of all the planets is named after the supreme Roman god Jupiter (known to the Greeks as Zeus). Much of the god's symbolism can be traced to the Mesopotamian Marduk, the patron god of Babylon. Marduk was the eldest son of Ea, the god of wisdom and lord of the sweet waters around the Earth. The Mesopotamian *Epic of Creation* (2nd millennium BCE) describes the birth of four generations of gods from the coupling of the salt-water ocean, Tiamat, and Apsu (the sweet waters incarnate), who vowed to kill his children. One child, Ea, slew Apsu instead.

Enraged, Tiamat called up a hideous brood of sea-monsters. Then Marduk, son of Ea, stepped forward and offered to defeat Tiamat if the gods would promise him supreme authority if he succeeded. So, the gods tested Marduk. They created a constellation, and asked him to destroy it and restore it with his will. Marduk did this and so was named the "herder of stars".

He destroyed Tiamat's monsters, and finally defeated Tiamat. He split her body into two, and made one half the sky and the other the earth. These acts established him as supreme creator sky-god, the title later given to the Greek Zeus. Greek myth parallels many elements of the story of Marduk, most impressively in the battle between the primeval gods and their offspring.

Zeus' rule gradually became more abstracted. Having been depicted as a lusty and capricious projection of human characteristics, from the 5th century BCE Zeus was reinterpreted by the Greek philosophers as an ultimate principle of divine order.

In the symbolism of the stars and planets, however, we are mostly offered an image of the original Zeus. His most obvious role is one of procreation, fathering a legion of gods and heroes from his sexual adventures. It is therefore appropriate that four of the planet's 16 moons are named after four of the god's lovers: Io, Europa, Ganymede and Callisto.

In Chinese astrology Jupiter was considered the divine law-maker, mirroring in the sky the "noble officials" on Earth.

A colour-enhanced photograph of Jupiter, which has an orbital cycle of approximately 12 years. It moves through one zodiac sign each year, and for this reason Chinese astrologers called it the "year star".

In Assyro-Babylonian myth, Saturn — the outermost of the seven "wandering stars" known to the ancients — represents the god Ninurta, brother of Nergal (see page 173). In a late version of the *Epic of Anzu* (7th century BCE), Saturn is identified as a planet of Fate, a correlation well known to later astrology. The wicked bird Anzu was envious of the power of the father of the gods, Enlil. In particular Anzu coveted the Tablets of Destiny, on which the fate of all beings is recorded. One day, while Enlil was bathing, Anzu grabbed the Tablets and flew with them to his remote lair. The Tablets gave their holder almost invincible power, and the gods were in despair at their loss. Wise Ea summoned the Earth-goddess to produce a hero-god from her own being, and she bore Ninurta. In the ensuing battle, Ninurta pierced Anzu through the heart and lungs and seized from him the Tablets. To honour his courage, Ninurta was appointed by the gods to be the guardian of the Tablets, and therefore, the overseer of fate.

The story of Cronus, the Greek Saturn, castrating his father Uranus is well known (see page 176). As Uranus lay dying, he prophesied that one of the sons of Cronus would in turn dethrone him. In order to avoid this, Cronus devoured his own children at the moment they were born, until he was tricked by Rhea, when she managed to smuggle out the baby Zeus, who later fulfilled the prophecy.

Saturn's identity as lord of both fate and time is a result of the Greek conflation of Cronus with the god Chronos (Time).

The absorption into Roman culture comes from the god Saturnus, originally an Italian agricultural deity, identified with an early king of Rome. His reign was so benevolent that it was considered a Golden Age. His festival, the Saturnalia, was a great annual celebration of the December solstice, when the Sun entered Capricorn, the zodiacal sign governed by Saturn. There was an orgy of feasting and merry-making, making this the pagan forerunner to our celebration of Christmas.

Saturn, with its famous rings, is the slowest-moving of all the planets, in keeping with its mythic association with age and time. The planet was associated too with the metal lead, a testament to its slow, heavy motion. In Chinese as well as European tradition, the god was personified as an old man.

Beyond the orbit of Saturn lie the so-called "modern" planets, too distant to be observed by the naked eye. Nevertheless, by naming these "new" planets after the gods, human imagination has remained faithful to the ancient symbolic attitude that created stellar mythology.

Uranus, twinkling at 6th magnitude, was discovered in 1781. Initially the new planet was named Herschel, after its discoverer, the English astronomer William Herschel (1738–1822), and the planet's sigil is a play on the letter "H" (see page 23). Herschel himself wished to name it *Georgium Sidus* ("George Star") after his patron George III. "Neptune" was also a strong contender, along with the horrible compromise *Neptune de George III*.

The planet Uranus shown juxtaposed against the surface of Miranda, the smallest of its five major satellites.

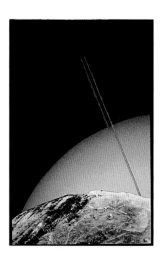

The name that finally won credence by the mid-19th century was originally proposed by the German astronomer Johann Bode (1747–1826). His suggestion of Uranus, the original sky-god and first father, seems in retrospect an obvious choice. It elegantly equates myth and science, because the order of the planets away from the Sun reflects the succession of the divine generations: Uranus was the father of Saturn (Cronus to the Greeks; see page 175), who was in turn the father of Jupiter (Zeus; see page 174).

The story of the god Uranus is among the most dramatic of classical themes. From primordial Chaos emerged the Earth-goddess Gaea, and from her was born the sky-god Uranus. He coupled with his mother to produce the hundred-handed giants, the one-eyed Cyclopes, and then the seven Titans. Uranus hated his offspring, and pushed them all into the underworld. Vengeful Gaea fashioned a flint sickle and instructed her newborn infant Cronus to castrate his father. Having done so, fatally, Cronus flung the genitals into the sea (see page 172).

Once Uranus is in place, it is possible to discern a poetic logic in the remaining planetary names. Neptune, the 8th-magnitude newcomer, discovered in 1846, was easily agreed. Of the ancestors of Jupiter, the only primordial divinity of the first rank missing from the planetary picture was Cybele or Ops (Rhea in Greek myth), but she lacked the authority of the ruling male line. The field was therefore clear for the dynasty founded by Jupiter, and especially his mighty brother Neptune (Poseidon), a rival to Jupiter in dignity, if not authority.

In Greek myth both Uranus as he died and Gaea, the Earth-goddess, prophesied that Cronus would be deposed by one of his own sons. This turned out to be Zeus, who led his siblings in a war against their father, and defeated and banished Cronus for ever.

The battle over, Zeus and his brothers Poseidon and Hades (Neptune and Pluto to the Romans) drew lots to decide which of them should rule the sky, which the sea, and which the underworld. Zeus gained the sky, Hades the underworld, and surly Poseidon was allotted the watery realm. Here he desired the sea-nymph Amphtrite, but she fled to the Atlas Mountains to escape. The messenger Delphinus won her back, and in gratitiude Poseidon placed him in the sky as a dolphin (see page 142).

Poseidon is an ancient Pelasgian deity, whose origins lie with the earliest Greeks. Horses were sacred to the god, and he sired the winged horse Pegasus (see pages 108–9); he kept white horses which we sometimes glimpse in the foaming crests of the waves.

The discovery of Pluto in 1930 was inspired by Percival Lowell (1865–1915), founder of the Flagstaff Observatory in Arizona. The name was suggested by Venetia Burney, an eight-year-old who knew her mythology. Her father immediately contacted an astronomy professor, who passed the idea on to the Royal Astronomical Society, and from there it went to Flagstaff. Curiously, the name reflects the initials of Percival Lowell: the sigil (see page 23) combines the letters P and L.

Remote Pluto is well named. In myth, Hades (Pluto's Greek equivalent, and the eldest brother of Zeus) gained the underworld in the selection by lots. This is the place of the dead, alien to all living things. Hades rarely visited the upper world, but he did emerge once to seize the beautiful Persephone (Proserpina) and drag her into his realm to become his queen.

An ancient Roman mosaic showing *The Triumph of Neptune*, god of the sea. Riding in a chariot, the god carries a trident, from which the planet's sigil is taken. Homer called Neptune "earth shaker", because when the god brandished his trident it was said to create earthquakes.

APPENDICES

THE ZODIAC SIGNS

The dates that the Sun enters each 30° section (sign) of the zodiac, and the degrees of celestial longitude at which each sign begins. Use this table to work out the position of the Sun on any date.

♈	Aries (0°) *March 21*	♎	Libra (180°) *September 23*
♉	Taurus (30°) *April 20*	♏	Scorpio (210°) *October 23*
♊	Gemini (60°) *May 21*	♐	Sagittarius (240°) *November 22*
♋	Cancer (90°) *June 22*	♑	Capricorn (270°) *December 22*
♌	Leo (120°) *July 23*	♒	Aquarius (300°) *January 20*
♍	Virgo (150°) *August 23*	♓	Pisces (330°) *February 19*

STAR TABLES: EPOCH 2000.0

This table lists 227 named stars. Given for each star is its designation (Des.), constellation (Con.; for key to constellation abbreviations, see page 27), apparent magnitude (Mag.), apparent magnitude and celestial longitude (Long.).

Star Name	Des.	Con.	Mag.	Long.
Acamar	θ	Eri	3.2	23°
Achernar	α	Eri	0.5	345°
Achird	η	Cas	3.4	40°
Acrab	β	Sco	2.6	243°
Acrux	α	Cru	1.4	221°
Acubens / Sartan	α	Cnc	4.3	133°
Adhafera	ζ	Leo	3.4	147°
Adhara	ε	CMa	1.5	110°
Adhil	ξ	And	4.8	37°
Ain	ε	Tau	3.5	68°
Aladfar	η	Lyr	4.4	300°
Albali	ε	Aqr	3.8	311°
Albireo	β	Cyg	3.1	301°
Alchiba	α	Crv	4.0	192°
Alcor	80	UMa	4.0	165°
Alcyone	25	Tau	2.9	60°
Aldebaran	α	Tau	0.9	69°
Alderamin	α	Cep	2.4	12°
Alfirk	β	Cep	3.2	35°
Algedi / Giedi	α	Cap	3.6	303°
Algenib	γ	Peg	2.8	9°
Algieba	γ	Leo	2.3	149°
Algol	β	Per	2.1	56°
Algorab	δ	Crv	3.0	193°

Star Name	Des.	Con.	Mag.	Long.
Alhena / Almeisan	γ	Gem	1.9	99°
Alioth	ε	UMa	1.8	158°
Alkaid / Benetnash	η	UMa	1.9	176°
Alkalurops	μ	Boö	4.3	213°
Alkes	α	Crt	4.1	173°
Almach	γ	And	2.2	44°
Alnair	α	Gru	1.7	315°
Alnasl	γ	Sgr	3.0	271°
Alnilam	ε	Ori	1.7	83°
Alnitak	ζ	Ori	1.8	84°
Alphard	α	Hya	2.0	147°
Alphecca / Gemma	α	CrB	2.2	222°
Alpheratz / Sirrah	α	And	2.1	14°
Alrischa	α	Psc	3.8	29°
Alshain	β	Aql	3.7	302°
Alsuhail	λ	Vel	2.2	161°
Altair	α	Aql	0.8	301°
Altais / Nodus II	δ	Dra	3.1	17°
Aludra	η	CMa	2.4	119°
Alula Australis	ξ	UMa	3.8	157°
Alula Borealis	ν	UMa	3.5	156°
Alya	θ	Ser	4.1	285°
Ancha	θ	Aqr	4.2	333°
Ankaa	α	Phe	2.4	345°
Antares	α	Sco	1.0	249°
Arcturus	α	Boö	0.0	204°
Arkab Posterior	β²	Sgr	2.5	285°
Arkab Prior	β¹	Sgr	3.9	285°
Arneb	α	Lep	2.6	81°
Arrakis	μ	Dra	4.9	234°
Ascella	ζ	Sgr	2.6	283°
Asellus Australis	δ	Cnc	3.9	128°
Asellus Borealis	γ	Cnc	4.7	127°
Aspidiske / Tureis	ι	Car	2.3	191°
Atik	o	Per	3.8	61°

Star Name	Des.	Con.	Mag.	Long.
Atria	α	TrA	1.9	260°
Avior	ε	Car	1.9	173°
Azha	η	Eri	3.9	38°
Azmidiske	ξ	Pup	3.3	126°
Baten Kaitos	ζ	Cet	3.7	21°
Beid	o¹	Eri	4.0	59°
Bellatrix	γ	Ori	1.6	80°
Betelgeuse	α	Ori	0.5	88°
Biham	θ	Peg	3.5	336°
Botein	δ	Ari	4.4	50°
Canopus	α	Car	−0.7	104°
Capella	α	Aur	0.1	81°
Caph	β	Cas	2.3	35°
Castor	α	Gem	1.6	110°
Cebalrai	β	Oph	2.8	265°
Chara	β	CVn	4.3	167°
Cih	γ	Cas	2.5	43°
Cor Caroli	α	CVn	2.9	174°
Coxa / Chort	θ	Leo	3.3	163°
Cursa	β	Eri	2.8	75°
Dabih	β	Cap	3.1	304°
Deneb	α	Cyg	1.3	335°
Deneb	ε	Del	4.0	314°
Deneb Algedi	δ	Cap	2.9	323°
Deneb Kaitos / Diphda	β	Cet	2.0	2°
Denebola	β	Leo	2.1	171°
Diadem	α	Com	4.3	188°
Dschubba	δ	Sco	2.3	242°
Dubhe	α	UMa	1.8	135°
Edasich	ι	Dra	3.3	184°
Elnath	β	Tau	1.7	82°
Eltanin / Etamin	γ	Dra	2.2	267°
Enif	ε	Peg	2.4	331°
Erakis	μ	Cep	4.1	9°
Errai	γ	Cep	3.2	60°

Star Name	Des.	Con.	Mag.	Long.
Fomalhaut	α	PsA	1.2	333°
Furud	ζ	CMa	3.0	97°
Gacrux	γ	Cru	1.6	216°
Giansar	λ	Dra	3.8	130°
Gienah	γ	Crv	2.6	190°
Gienah	ε	Cyg	2.5	327°
Gomeisa	β	CMi	2.9	112°
Grumium	ξ	Dra	3.8	264°
Hadar / Agena	β	Cen	0.6	233°
Hamal	α	Ari	2.0	37°
Hassaleh	ι	Aur	2.7	76°
Heze	ζ	Vir	3.4	202°
Hoedus I	ζ	Aur	3.8	78°
Hoedus II	η	Aur	3.2	79°
Homam	ζ	Peg	3.4	346°
Hyadem I	γ	Tau	3.6	65°
Hyadem II	δ	Tau	3.8	66°
Izar / Pulcherrima	ε	Boö	2.4	208°
Kaffaljidmah	γ	Cet	3.5	39°
Kajam/Cujam	ω	Her	4.6	241°
Kaus Australis	ε	Sgr	1.9	275°
Kaus Borealis	λ	Sgr	2.8	276°
Kaus Meridionalis	δ	Sgr	2.7	274°
Keid	o²	Eri	4.4	60°
Kitalpha	α	Equ	3.9	323°
Kochab	β	UMi	2.1	133°
Kornephoros	β	Her	2.8	241°05
Kraz	β	Crv	2.7	197°
Kuma	ν	Dra	4.9	250°
Lesath	υ	Sco	2.7	256°
Maasym	λ	Her	4.4	259°
Maaz / Almaaz	ε	Aur	3.0	78°
Marfik	λ	Oph	3.8	245°
Markab	α	Peg	2.5	353°
Markeb	κ	Pup	4.7	123°

Star Name	Des.	Con.	Mag.	Long.
Matar	η	Peg	2.9	355°
Mebsuta	ε	Gem	3.0	99°
Megrez	δ	UMa	3.3	151°
Meissa / Heka	λ	Ori	3.4	83°
Mekbuda	ζ	Gem	3.8	104°
Menkalinan	β	Aur	1.9	89°
Menkar	α	Cet	2.5	44°
Menket	θ	Cen	2.1	222°
Merak	β	UMa	2.4	139°
Mesarthim	γ	Ari	4.6	33°
Miaplacidus	β	Car	1.7	211°
Mimosa	β	Cru	1.3	221°
Minelauva	δ	Vir	3.4	191°
Minkar	ε	Crv	3.0	191°
Mintaka	δ	Ori	2.2	82°
Mira	ο	Cet	3.0	31°
Mirach	β	And	2.1	30°
Mirfak / Algenib	α	Per	1.8	62°
Mirzam	β	CMa	2.0	97°
Mizar	ζ	UMa	2.3	165°
Muphrid	η	Boö	2.7	199°
Naos	ζ	Pup	2.3	138°
Nashira	γ	Cap	3.7	321°
Nekkar	β	Boö	3.5	204°
Nihal	β	Lep	2.8	79°
Nunki	σ	Sgr	2.0	282°
Nusaken	β	CrB	3.7	219°
Peacock	α	Pav	1.9	293°
Phact	α	Col	2.6	82°
Phecda	γ	UMa	2.4	150°
Pherkad	γ	UMi	3.1	141°
Polaris / Cynosaura	α	UMi	2.0	88°
Pollux	β	Gem	1.1	113°
Porrima	γ	Vir	2.8	190°
Praecipua	46	LMi	3.8	150°

Star Name	Des.	Con.	Mag.	Long.
Procyon	α	CMi	0.4	115°
Rana	δ	Eri	3.5	50°
Ras Algethi	α	Her	3.2	256°
Ras Alhague	α	Oph	2.1	262°
Ras Elased Australis	ε	Leo	3.0	140°
Ras Elased Borealis	μ	Leo	3.9	141°
Rastaban / Alwaid	β	Dra	2.8	251°
Regulus / Cor Leonis	α	Leo	1.4	149°
Rigel / Algebar	β	Ori	0.1	76°
Rigel Kentaurus / Toliman	α	Cen	−0.3	239°
Rotanev	β	Del	3.5	316°
Ruchbah / Ksora	δ	Cas	2.7	47°
Rukbat / Alrami	α	Sgr	4.0	286°
Sabik	η	Oph	2.4	257°
Sadachbia	γ	Aqr	3.8	336°
Sadalbari	μ	Peg	3.5	354°
Sadalmelik	α	Aqr	3.0	332°
Sadalsuud	β	Aqr	2.9	323°
Sadr	γ	Cyg	2.2	324°
Saiph	κ	Ori	2.1	86°
Sargas	θ	Sco	1.9	265°
Sarin	δ	Her	3.1	254°
Sceptrum	53	Eri	3.9	65°
Scheat	β	Peg	2.4	359°
Schedar	α	Cas	2.2	37°
Segin	ε	Cas	3.4	54°
Seginus	γ	Boö	3.0	197°
Shaula	λ	Sco	1.6	264°
Sheliak	β	Lyr	3.5	288°
Sheratan	β	Ari	2.6	33°
Sirius	α	CMa	−1.5	104°
Situla	κ	Aqr	5.0	339°
Skat	δ	Aqr	3.3	338°
Spica	α	Vir	1.0	203°
Sualocin	α	Del	3.8	317°

Star Name	Des.	Con.	Mag.	Long.
Subra	ο	Leo	3.5	144°
Sulaphat	γ	Lyr	3.2	291°
Syrma	ι	Vir	4.1	213°
Tabit	π³	Ori	3.2	71°
Talitha	ι	UMa	3.1	122°
Tania Australis	μ	UMa	3.1	141°
Tania Borealis	λ	UMa	3.5	139°
Tarazed / Reda	γ	Aql	2.7	300°
Tejat Posterior	μ	Gem	2.9	95°
Tejat Prior / Propus	η	Gem	3.3	93°
Theemin	ν²	Eri	3.8	59°
Thuban	α	Dra	3.7	157°
Unukalhai	α	Ser	2.7	232°
Vega	α	Lyr	0.0	285°
Vindemiatrix	ε	Vir	2.8	189°
Wasat	δ	Gem	3.5	108°
Wasn	β	Col	3.1	86°
Wezen	δ	CMa	1.9	113°
Yed Posterior	ε	Oph	3.2	243°
Yed Prior	δ	Oph	2.7	242°
Zania	η	Vir	3.9	184°
Zaurak	γ	Eri	3.0	53°
Zavijava	β	Vir	3.6	177°
Zibal	ζ	Eri	4.8	43°
Zosma	δ	Leo	2.6	161°
Zuben Elakrab	γ	Lib	3.9	235°
Zuben Elgenubi	α	Lib	2.8	225°
Zuben Eschamali	β	Lib	2.6	229°

GLOSSARY

Words in SMALL CAPITALS have their own entries in the glossary.

0° ARIES The first point of the ZODIAC sign Aries, where the ECLIPTIC and CELESTIAL EQUATOR intersect. It is the point at which all equatorial and ecliptic measurements on the CELESTIAL SPHERE begin.

ASTERISM A notable grouping of stars.

BAYER DESIGNATION The modern naming of principal stars in each constellation by letters of the Greek alphabet. The system was formed by Johann Bayer (1572–1625), who, in 1603, published the first star atlas to cover the whole sky. Usually, but not always, α is the LUCIDA.

BINARY A double star, made up of two associated stars orbiting around a mutual centre of gravity. This often produces the effect of a VARIABLE star. "Visual binaries" are unconnected stars (ie. they have no shared centre of gravity) that appear close by in line of sight.

CELESTIAL EQUATOR A projection of the Earth's equator, midway between the poles, onto the CELESTIAL SPHERE.

CELESTIAL LONGITUDE A measure along the ecliptic from 0° to 360°. Zodiacal longitude is the same measure divided into twelve 30° signs of the ZODIAC.

CELESTIAL POLES The points, north and south, at which the Earth's poles, if projected, would cut the CELESTIAL SPHERE.

CELESTIAL SPHERE A projection of the Earth's surface into the sky. It is used by astronomers to map the "fixed stars" and to make celestial measurements.

CIRCUMPOLAR Stars that do not appear to set from a given geographic latitude. The term is used in a general sense for stars with a DECLINATION greater than $66\frac{1}{2}°$N or S.

COMET A small conglomeration of dust and gas, which orbits the Sun. Comets often have irregular or highly elliptical orbits. Their gases light up as they approach the Sun.

CULMINATION The passage of a star across the observer's MERIDIAN. Unless specified, "upper culmination" is understood to be closer to the observer's ZENITH. "Lower culmination" is the passage farther from the observer's zenith, out of sight unless CIRCUMPOLAR at that geographic latitiude.

DECLINATION A measure in degrees north (+) or south (–) of the CELESTIAL EQUATOR.

ECLIPTIC The Sun's apparent track through the star-fields, marking the Earth–Sun orbital plane. The band of the ZODIAC is mapped along this path.

EQUINOCTIAL COLURE The MERIDIAN (north–south) line that passes through the EQUINOX POINTS.

EQUINOX POINT One of two intersections of the CELESTIAL EQUATOR and the ECLIPTIC, crossed by the Sun on March 21 on one side of the CELESTIAL SPHERE, and on September 23 directly opposite on the other side of the Sphere. At the equinoxes day and night are of equal length all over the world. The March equinox point (0° ARIES) is the start of the equatorial and ECLIPTIC measures, and of the band of the ZODIAC.

ELONGATION The angular distance between the Sun, and the Moon or a planet.

FLAMSTEED NUMBER John Flamsteed (1646–1719) catalogued and numbered 2,935 stars, complementing the work of BAYER.

GALAXY A stellar system, in which millions of stars are held together by gravity. Our Galaxy is the Milky Way — a spiral of stars, dust and gas, twisting out from its centre in the constellation Sagittarius.

GLOBULAR CLUSTER A cluster of between 100,000 and one million stars, sharing the same gravitational field.

HELIACAL RISING/SETTING A star's, planet's or constellation's rising just before the Sun (morning star) or setting just after the Sun (evening star).

HERMETIC STAR One of 15 key stars employed in medieval European and Renaissance magic. They are (with their constellations in brackets): Alcyone (Pleiades; Tau), Aldebaran (Tau), Algol (Per), Alphecca (CrB), Antares (Sco), Arcturus (Boö), Benetnash (UMa), Capella (Aur), Deneb Algedi (Cap), Gienah (Crv), Procyon (CMi), Regulus (Leo), Sirius (CMa), Spica (Vir), Vega (Lyr). In this book their SIGILS are depicted as a background underlying the text for the relevant constellation.

HORIZON The circular plane below which stellar observation for a specific geographic latitude is impossible.

LUCIDA The brightest star in a constellation. This is usually, but not always, given the BAYER DESIGNATION α.

MAGNITUDE A measure of the luminosity of a star or planet at source (absolute magnitude) or as seen from Earth (apparent magnitude). Unless otherwise stated, magnitude figures given this book are apparent magnitudes.

MERIDIAN A north–south line on the CELESTIAL SPHERE; also, the great circle passing through the observer's ZENITH and the Celestial North and South Poles.

METEOR Debris left behind, usually by a comet, burning as it enters the Earth's atmosphere. Lumps that reach the Earth's surface are "meteorites".

NEBULA A luminous cloud of interstellar dust or gas. A planetary nebula is so-called because it appears to have rings as do some planets.

NADIR The point on the CELESTIAL SPHERE directly beneath the observer (opposite the ZENITH).

PRECESSION OF THE EQUINOXES The slow backward motion of the EQUINOX points against the fixed star-fields in a 25,868-year cycle. It occurs as a result of the slow wobble of the Earth's polar axis.

RIGHT ASCENSION (RA) A measure along the celestial equator in hourly sections from 0 to 24 hours, beginning at the March EQUINOX point.

SIGIL A derivation from an ancient glyph, or a newly-created symbol, to denote a planet, star or constellation.

SOLSTICE The point on the ecliptic where the Sun reaches its maximum DECLINATION south, at 90° CELESTIAL LONGITUDE (June 22) or north at 270° (December 22).

SOLSTICIAL COLURE The MERIDIAN (north–south) line that passes through the SOLSTICE points.

SUPERNOVA An exploding star that pours out light and energy. On rare occasions supernovae appear like brilliant new stars, which rapidly fade from view.

TROPICS The region on the Earth between the Tropic of Cancer ($23\frac{1}{2}$°N) and the Tropic of Capricorn ($23\frac{1}{2}$°S). The Sun is directly overhead at these geographic latitudes at the June (Cancer) and December (Capricorn) solstices.

VARIABLE A star that varies in its magnitude regularly or irregularly.

ZENITH The point on the CELESTIAL SPHERE directly overhead.

ZODIAC The apparent pathway of the Sun, Moon and planets, divided into 12 equal (30°) Signs. It is defined by the Moon's maximum latitude (its maximum swing) 6° above or below the ECLIPTIC. It has also been described as 9° either side (the maximum latitude of the visible planets). The zodiac signs must be distinguished from the zodiac constellations, which occupy irregularly-sized portions along the ecliptic.

BIBLIOGRAPHY

Allen, Richard Hinckley   *Star Names: Their Lore and Meaning*, New York: Dover (1963)

Aratus   "Phaenomena", *Callimarchus, Lycrophon and Aratus* (tr. A.W. Mair and G.W. Mair), Loeb 129, Cambridge, Massachusetts: Harvard University Press, and London: Heinemann (1989)

Bakich, Michael E.   *The Cambridge Guide to the Constellations*, England: Cam. UP (1995)

Cornelius, Geoffrey and Devereux, Paul   *The Language of the Stars and Planets*, London: Duncan Baird Publishers, and San Francisco: Chronicle Books (2003)

Dibbon-Smith, Richard   *Starlist 2000*, New York: Wiley Science Editions (1992)

Ellyard, David and Tirion, Wil   *The Southern Sky Guide*, England: Cambridge UP (2001)

Graves, Robert   *The Greek Myths*, England: Penguin Combined Edition (1992)

Hirshfield, A. and Sinnott, R. W.   (ed.) *Sky Catalogue 2000.0* (Vol. 1), England: Cambridge University Press (1982)

Manilius   *Astronomica* (tr. G.P. Goold), Loeb Classics 489, Cambridge, Massachusetts: Harvard University Press, and London: Heinemann (1977)

Ridpath, Ian and Tirion, Wil   *Stars and Planets*, London: HarperCollins (2nd ed., 1995)

Robson, Vivian E.   *The Fixed Stars and Constellations in Astrology*, London: Cecil Palmer (2003)

Staal, Julius D.W.   *The New Patterns in the Sky*, Virginia: MacDonald and Woodward (1998)

Tirion, Wil   *Cambridge Star Atlas*, England: Cambridge University Press (2001)

PICTURE CREDITS

Abbreviations:
AKG: Archiv für Kunst und Geschichte;
BAL: Bridgeman Art Library;
CWC: Charles Walker Collection;
MEPL: Mary Evans Picture Library;
SPL: Science Photo Library

The Author and the Publisher would like to thank each of the following for their kind permission to reproduce their photographs in this book:

Page 7: BAL/Musée Condé, Chantilly; 8: BAL; 10: AKG; 10: Fine Art Photographic Library; 13: Images/CWC; 14: Magie Hyde/ *Jung and Astrology*; 17: Images/CWC; 22: BAL; 24: SPL; 25: SPL/NASA; 26: e.t. archive/University Library, Istanbul; 40: National Maritime Museum; 40: Images; 45: BAL/British Library; 49: British Library/Ms. Harley 64/9 7 f2 v.; 51: BAL/British Library; 61: Robert Harding Picture Library; 65: Bodleian Library; 68: BAL/British Library; 69: BAL/British Library; 75: Images/ CWC; 77: British Museum; 79: BAL/British Library; 85: BAL/Duomo, Florence; 87: BAL; 89: Images/CWC; 93: Bodleian Library; 95: British Museum; 97: Images/CWC; 99: British Museum; 111: Bodleian Library MS. Marsh 144 p111; 113: Bodleian Library; 117: BAL/Bibliothèque Nationale, Paris; 118: Werner Forman Archive; 31: MEPL; 122: British Library Ms. Harley 647 f4 v.; 123: MEPL; 131: e.t. archive/Carthage Museum; 134: BAL; 134: Images/CWC; 166: The National Library of Copenhagen; 166: AKG; 168: e.t. archive; 169: e.t. archive; 170: e.t. archive; 172: Images/CWC; 173: Images/CWC; 174: SPL/ NASA; 175: SPL/NASA; 176: SPL/NASA; 177: AKG; 178: AKG.

INDEX

Note: page numbers in **bold** refer to information contained in captions; page numbers in *italics* refer to information contained in tables.

N

Nabu 171
nadir 15
Naiads 91
Namtar 173
NASA 66
Nashira 64, **65**
Naxos 82
nebulae 25
Nekkar 52, **53**
Nemean lion 94, 98
Nephele 48, 75
Neptune (planet) 167, 176-7
Nereids 70-1
Nereus 70, 88
Nergal 117, 173, 175
Nessus 95
New York 26
New Zealand **118**
NGC 86 (Double Cluster) 110
NGC 104 (47 Tucanae) 25, 164
NGC 2070 (Tarantula Nebula) 142
NGC 2237 and 2244
    (Rosette Nebula) 150
NGC 2264 (Cone Nebula) 150
NGC 4755 (Jewel Box) 84
NGC 5139 (Centauri) 25, **25**
NGC 6193 137
NGC 6205 (M13) 25
NGC 7000 (North
    American Nebula) 86
NGC 7293 (Helix Nebula) 44
night, length of 17
night sky 6, 11-39
Nihal 147, **147**
Nile 58-9, 74, 90, 98, 107,
    115, 132, 168-9
Nine Muses 109
Ninurta 175
Noah's Ark 66, 140
Norma (Nor) 151, **151**
North Celestial Pole 15, **15**,
    88, 89, 112, 128, 140

North Pole 16, 52, 70, 124, 126, **127**
North Wind 50
Northern Cross 31, 86
Northern hemisphere 6,
    10, 25, 27-9, **40**
January 28, *28*, **28**
July 31, *31*, **31**
major constellations 41
March 29, *29*, **29**
May 30, *30*, **30**
November 33, *33*, **33**
September 32, *32*, **32**
November 33, *33*, **33**, 39, *39*, **39**
Nunki 116, **117**

O

Oannes 64
Oceanid 90
Octans (Oct) 139, 145, 151, **151**
Oenomaus, King 50-1
Olympians 89
Olympus 45, 80, 95, 137, 172
omen-texts 12
Ophiuchus (Oph) 37-8, **37-8**,
    104-5, **105**, 118-19, 158
Ops 176
see also Rhea
oracle of Ammon 70, 72
oracle at Delphi 48, 94, 168
oracle of Zeus, Dodona 67
Orion Nebula 25
Orion (Ori) 28-9, **28-9**, 33-5,
    33-5, 39, **39**, 58-60, **60-1**,
    62, **91**, 92, 106-7, **107**,
    119, 120, 122, 147, 150
Orpheus 102
Osiris 59, **61**, 107, 115, 121
Ovid 60, 92, 98, 123
Oxyrhynchus 115

P

pagans 171, 175

Palestine 42, 43, 70
Pan 64-5
Papin, Denis 136
Paris (Trojan prince) 172
Pavo (Pav) 37, **37**, 121, 152, **152**, 159
Peacock 37, 152, **152**, 159, **159**
Pegasus (Peg) 32, 38, **38**, 42, **42**,
    44, 72, **73**, 78, **79**, 108-9, **109**, 177
Great Square of 32, 38, 42,
    70, 72, 108, **109**, 112, **115**
Pelias 67
Pelops 50-1
Perseids meteor shower 110
Persephone 131, 160, 177
Perseus (Per) 33, **33**, 42-3, 71-2,
    73, 78, **79**, 109-11, **111**, 138
Persian Gulf 64
Persian tradition 12, 41,
    92, 121, 156, 169
Phact 140, **140**
Phaethon 90-1, 168
Phecda 124, **125**
Philira 74
Phineas 68, 71
Phoenician tradition 92, 128
Phoenix (Phe) 38, **38**, 153, **153**
Phrixus 48-9, 66-7
Pictor (Pic) 153, **153**
Pisces (Psc) 18, **18-19**, 32, **32**, 72,
    73, 78, 112-14, **113**, 116, 180, **180**
Piscis Austrinus (PsA) 32, **32**, 38-9,
    38-9, 44, 114-16, **115**, **134**, 144, 157
Plancius, Petrus 138, 140, 150
planetary nebulae 25
planets 22-3
see also specific planets
planisphere 26-7
Pleiades 25, 39, 120, 121, 122, **122**, **123**
Pleione 122
Pliny 123
Plough (Big Dipper) 12, **14**, **16**,
    29-30, **29-30**, **53**, 98, 106,
    124, **125**, 126, **127**, 128, 130